Mathematics for Dynamic Modeling

Mathematics for Dynamic Modeling

Edward Beltrami

Department of Applied Mathematics
State University of New York at Stony Brook
Stony Brook, New York

ACADEMIC PRESS, INC.
Harcourt Brace Jovanovich, Publishers
Boston San Diego New York Berkeley
London Sydney Tokyo Toronto

ACADEMIC PRESS, INC.
San Diego, California 92101

United Kingdom Edition published by
ACADEMIC PRESS LIMITED
24-28 Oval Road, London NW1 7DX

Library of Congress Cataloging-in-Publication Data

Beltrami, Edward J.
Mathematics for dynamic modeling.

Bibliography: p.
Includes index.
1. Dynamics — Mathematical models. 2. Differentiable dynamical systems. 3. Mathematical models. I. Title.
QA845.B53 1987 003 87-1833
ISBN 0-12-085555-0 (alk. paper)

Cover design by Paola Di Stefano
Photograph by Bruce T. Martin © 1987

PRINTED IN THE UNITED STATES OF AMERICA
88 89 90 91 9 8 7 6 5 4 3 2

To Barbara,
fide et amore

Contents

Preface

> This model will be a simplification and an idealization, and consequently a falsification. It is to be hoped that the features retained for discussion are those of greatest importance in the present state of knowledge.
>
> —A. M. Turing, "The Chemical Basis of Morphogenesis", *Philos. Trans. Royal Soc.*, London, Ser. B, Vol. 237, 1952, 37–72.

One of the features that distinguishes applied mathematics is its interest in framing important questions about the observed world in a mathematical way. This process of translation into a mathematical form can give a better handle for certain problems than would be otherwise possible. We call this the modeling process. It combines formal reasoning with intuitive insights. Understanding the models devised by others is a first step in learning some of the skills involved, and that is how we proceed in this text, which is an informal introduction to the mathematics of dynamical systems. The mathematical formulations are in terms of linear and nonlinear differential equations, ordinary and partial, in at most three dependent variables, and some difference equations. The problems illustrate the concepts of equilibrium and stability, bifurcation, limit cycles, and chaos. Among the topics treated in an elementary manner are reaction–diffusion and shock phenomena, Hopf bifurcations in $\mathbf{R}^3$, cusp catastrophes, and strange attractors. Applications are to the temporal and spatial dynamics of interacting populations, geophysical and physiological models, and oscillatory mechanical systems.

This book is intended as a text for upper undergraduate and first-year graduate courses in mathematical modeling or for an applications-oriented second course in differential equations as generally offered in departments of applied mathematics, systems science and engineering, and in the biological sciences. However, because of the importance of these ideas in contemporary applied mathematics the accessible presentation in this book should be of interest to a number of non-specialists in related fields who wish to make contact with a rapidly growing area of research.

A number of central results are motivated and explained by a combination of a little rigor and a lot of intuition. This permits us to give elementary discussions of sophisticated models. For example, the usual treatment of nonlinear dynamical systems, at least in an undergraduate course, is restricted to a study of equilibria and cycles. Our book provides a glimpse into other, more complicated dynamical behavior, at a level compatible with the background of upper undergraduate or first-year graduate students.

The reader will not find the usual "theorem–proof" apparatus of most mathematics books. The present text aims for a less formal presentation in keeping with the way that applied mathematics is usually practiced. Prerequisites for reading the book are modest. The student should understand the basic elements of differential equations and matrix theory as generally presented in an undergraduate calculus course. Some of the models require, additionally, a knowledge of first-year physics concepts. These can be skipped, however, if the main interest is biological and physiological modeling since there is ample material to satisfy a variety of needs. The Appendix reviews the few basic facts about ordinary differential equations that are needed, with all proofs in $\mathbf{R}^2$ for simplicity.

The first four chapters, Part 1 of the book, introduce the reader to the notions of equilibrium and stability in differential equation modeling. These occur in the guise of simple models in the plane. Important tools are discussed including linearization, phase plane methods, feedback concepts, the use of Liapunov functions, and an excursion into the question of model robustness. With few exceptions, the material is fairly standard and is accessible to prepared undergraduates.

There are nonlinear models in which the limiting behavior of orbits can be more complicated. This is where our attention is focused in Chapters Five through Nine, Part 2 of the book. Here we enter a realm in which much remains to be done and where work is still being actively carried out. Our presentation is therefore tentative, and should only be thought of as a guide to some of the more interesting modeling possibilities. The level of presentation is slightly higher than in the preceding chapters and requires more maturity on the part of the student even though the mathematical requirements remain virtually the same. By way of reward, the examples are more interesting and less familiar than those in the first four chapters.

Chapter Five is an introduction to partial differential equation models, linear and nonlinear. The problems illustrate important ideas such as traveling wave solutions, shock propagation, and the growth of stable inhomogeneous spatial patterns in the presence of diffusion. No prior exposure to these ideas is assumed. Chapters Six and Seven treat models that lead in a natural way to cyclic behavior, bifurcation phenomena, and cusp catastrophes. Chapter Eight is an introduction to strange attractors and chaos. The last chapter introduces models that are based on the notion of optimization. These were included in this book because they lead to differential equations whose orbit behavior can be understood in terms of ideas introduced earlier.

Each chapter includes exercises, some of which expand on the material in the text. The reader will find sources for the models discussed in the "References And a Guide To Further Readings" at the end of the book. Here one can also catch up with specific references to theorems left unproven in the text and on other mathematical details.

The specific "real world" examples in the book are chosen from questions that have been asked in engineering, geophysics, physiology, and ecology. What this teaches about physiology and geophysics is incidental. One does not have to know much about these topics to begin with nor will the student qualify as an expert on them when he or she finishes. Many other applications could have been chosen to illustrate the modeling ideas, but we hope the students will enjoy those that were picked.

The essence of modeling, as we see it, is that one begins with a nontrivial word problem about the world around us. We then grapple with the not always obvious problem of how it can be posed as a mathematical question. Emphasis is on the evolution of a roughly conceived idea into a more abstract but manageable form in which inessentials have been eliminated. One of the lessons learned is that there is no best model, only better ones.

The model is only a suggestive metaphor, a fiction about the messy and unwieldy observations of the real world. In order for it to be persuasive, to convey a sense of credibility, it is important that it not be too complicated and that the assumptions that are made be clearly in evidence. In short, the model must be simple, transparent, and verifiable. We have allowed ourselves to be guided by these precepts in choosing the examples.

The book has been used in one-semester courses at both a junior–senior undergraduate level as well as a first-year graduate course. Most of the students have been majors in applied mathematics, but the course also included a considerable number of engineering and biology students with a quantitative bent. We are aware, of course, that there is also another kind of reader, the kind who dips into books out of curiosity or for amusement. Many of our examples were chosen with this person in mind.

Conspicuously absent is a discussion of specific computer codes for nonlinear differential equations. We strongly encourage that an instructor augment the modeling course with numerical work that illustrates some of the long-term behavior that may be expected. This is especially true in Chapter Eight, where computer experimentation is not only an asset but a necessity if one is to grasp what is going on.

I wish to thank my many students in AMS 331 and AMS 500 at the State University of New York at Stony Brook as well as those in my CNR summer course in Cortona, Italy, for their forebearance in using several preliminary and garbled versions of the notes that led to this book. Professors Karl Hadeler of the Biology Institute at the University of Tübingen and Courtney Coleman of Harvey Mudd College in Claremont read portions of the manuscript and made very helpful comments which led to substantial improvements. Bill Sribney, my editor at Academic Press, also deserves special thanks for his thorough review of the text. His meticulous editing greatly enhanced the book's clarity and readability.

During the final revision, I also benefited from the gracious hospitality of G. Stianti and C. Mascheroni at the Castello di Volpaia. The serene and lovely vistas from this tiny medieval hamlet, perched high in the Tuscan hills, were inspirational. To Giovannella and Carlo, dedicated conservationists of a precious Tuscan culture and producers of a splendid wine, I offer my warmest thanks.

Part 1

First Thoughts on Equilibria and Stability

Chapter One

Simple Dynamic Models

1.1 Back and Forth, Up and Down

Newton's laws of motion, formulated several centuries ago, provide some of the earliest examples of mathematical modeling. Consider a body having a *constant mass* m that is concentrated at a point and whose motion is restricted to a single line. The body accelerates under the influence of a force. Otherwise, it is either at rest or in motion at a uniform velocity. If the line of motion is the x axis, then its position, velocity, and acceleration at any time t are denoted by $x(t)$, $\dot{x}(t)$, and $\ddot{x}(t)$ respectively. One of Newton's laws is that if F is the sum of the forces acting on the body, its motion satisfies the differential equation

$$F = m\ddot{x}. \tag{1.1}$$

Several simple examples have been chosen to illustrate specific instances of 1.1. All of them lead to linear second-order differential equations.

Example 1.1 Consider a body of mass m lying on a very smooth table along the x axis. It is tethered to a spring that is itself attached to a wall. The position of rest for the *mass–spring system* is at $x = 0$. Here the spring is neither stretched nor compressed. If the mass is now pulled a little to the right or pushed a bit to the left and then released, we would expect the stretched or compressed spring to behave as a force that tends to restore the mass back to its position of rest. It is convenient to postulate an ideal spring, namely one that is massless and that gives rise to a restoring force proportional to the distance the body is displaced from rest. Real springs, if not distorted too much, appear to behave closely like the ideal one. Letting k denote the constant of proportionality, we can write down the equation that governs the motion of the mass–spring system by utilizing 1.1:

$$m\ddot{x} = -kx. \tag{1.2}$$

Note that the negative sign in 1.2 is appropriate because if $x > 0$ then the spring is stretched and so the restoring force is to the left, or in the negative direction. The converse is true when $x < 0$.

Example 1.2 Let us now remove the modeling assumption of a smooth table top. In actuality, there is always some resistance to motion because of friction. Resistance is a force that operates in a direction opposite to motion. Several formulations of this force are possible. One of these is called *viscous damping* and is associated with motion through a fluid medium such as air. Resistance in this case is known to be roughly proportional to the velocity and can be expressed as $-k_1\dot{x}$ for a suitable constant k_1.

In the case of our horizontal mass–spring system, the viscous effect of air resistance is less important than table top roughness. When the body is displaced from rest and let go, it adheres to the surface and there is no movement unless the restoring force is large enough. This means that for $\dot{x} = 0$ the frictional force exactly counterbalances the restoring force until, with sufficient stretching or compression, the restoring force eventually dislodges the body from rest. Once the body begins to move, the frictional force is roughly constant with a direction which depends on the velocity. This resistive force is known as *dry* or *Coulomb* friction and is defined for $\dot{x} \neq 0$ by $-k_1 \operatorname{sgn} \dot{x}$. In either situation, the inclusion of resistance acts to oppose motion and an application of 1.1 now gives one or the other of the following differential equation models, depending on the assumptions made:

$$m\ddot{x} = -kx - k_1\dot{x} \quad \text{or} \tag{1.3}$$

$$m\ddot{x} = -kx - k_1 \operatorname{sgn} \dot{x}, \qquad \dot{x} \neq 0. \tag{1.4}$$

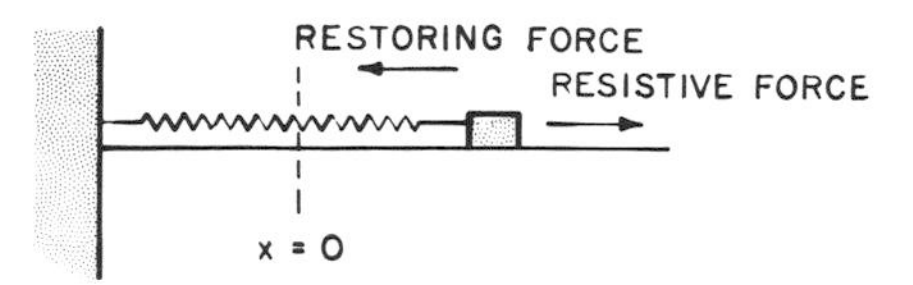

Figure 1.1 The horizontal mass–spring system.

The negative signs in the resistive terms are again appropriate because if $\dot{x} > 0$, so that motion is to the right, then resistance should be to the left or in the negative direction. The opposite is true when $\dot{x} < 0$ (Figure 1.1).

Example 1.3 We can introduce a new modeling assumption by allowing the mass–spring system to be suspended vertically from a beam. Motion is now along the y axis (the positive direction is upward) and the zero position corresponds to the rest position of the massless spring when the body also has zero mass. However, when mass is nonzero, a gravitational force comes into play, which tends to pull the body downward. Since the spring is thereby stretched, it provides a restoring force upward. At some point the two opposing forces balance each other and the system is again at rest, but this time at some lower position $y_0 < 0$ along the y axis. What the value of y_0 actually is will be decided on momentarily. Our new assumption is that the gravitational force per unit mass is a constant g. Since air resistance is a form of viscous damping, Equation 1.1 becomes the following statement for movement along the vertical y axis:

$$m\ddot{y} = -ky - k_1\dot{y} - mg. \tag{1.5}$$

The sign of the gravitational force is negative since it always acts in the negative y axis direction (Figure 1.2).

It is useful to rewrite 1.5 by dividing through by m and define $r = k_1/m$ and $\omega^2 = k/m$. Then

$$\ddot{y} + r\dot{y} + \omega^2 y = -g. \tag{1.6}$$

Letting $z = y + g/\omega^2$, it is readily seen that

$$\ddot{z} + r\dot{z} + \omega^2 z = 0, \tag{1.7}$$

which is identical to Equation 1.3 with z in place of x. The rest position for a horizontal mass–spring system is at the origin, and therefore the

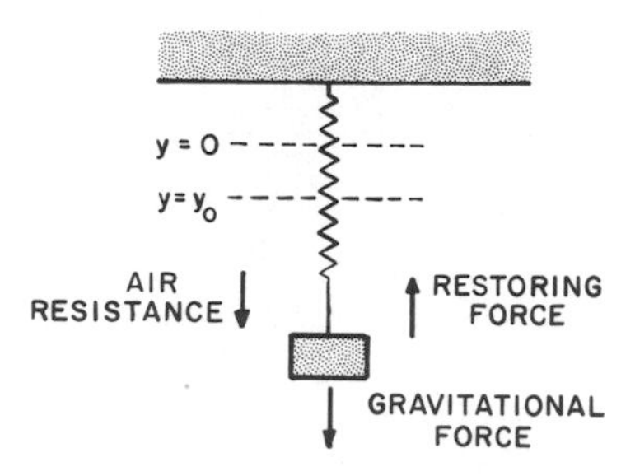

Figure 1.2 The vertical mass–spring system.

same is true for 1.7. But $z = 0$ is equivalent to $y = -g/\omega^2$, and so the rest position in a vertical direction, where gravitational pull has an effect, is now at $y_0 = -g/\omega^2 < 0$.

1.2 The Harmonic Oscillator

It is well known (see Appendix) that the general solution to Equation 1.7 can be obtained as

$$z(t) = c_1 e^{\lambda_1 t} + c_2 e^{\lambda_2 t}, \tag{1.8}$$

where c_1, c_2 are constants to be determined and λ_1, λ_2 are complex numbers that satisfy the algebraic equation

$$\lambda^2 + r\lambda + \omega^2 = 0. \tag{1.9}$$

Relation 1.9 is assumed to have two distinct roots. From a modeling viewpoint, this is the most typical or, mathematically speaking, the generic case. What is meant by this will be elaborated on later, but essentially it means that only under exceptional circumstances are the roots equal. They may be written as

$$\lambda_i = -\frac{r}{2} \pm \sqrt{\frac{r^2}{4} - \omega^2}.$$

Consider first the case in which there is only a little damping. We interpret this to mean that $r^2/4 < \omega^2$. Then the roots are complex and

1.8 becomes

$$z(t) = e^{-\sigma t}(c_1 \cos qt + c_2 \sin qt), \tag{1.10}$$

where $\sigma = r/2$ and $q = \sqrt{\omega^2 - r^2/4}$. If the reader is not clear how this comes about, we suggest he or she read the Appendix before continuing.

Consider first the case in which $r = 0$ (no resistance). Then $\sigma = 0$ and 1.10 shows that

$$z(t) = c_1 \cos \omega t + c_2 \sin \omega t. \tag{1.11}$$

But $z(0) = z_0$, for some $z_0 \neq 0$ (since the body is initially stretched or compressed) and $\dot{z}(0) = 0$ (because the body has no velocity when it is first let go). Applying these conditions to 1.11 allows us to conclude that

$$z(t) = z_0 \cos \omega t, \tag{1.12}$$

and so the mass oscillates back and forth about the zero position with a *period* T (namely the time for one complete cycle) determined by $\omega T = 2\pi$. Therefore the period is $T = 2\pi/\omega$. The quantity ω is called the *frequency of oscillation* and it represents the number of cycles which are completed during an interval 2π. T decreases as the frequency ω increases. The value of $z(0)$ represents the *amplitude*, namely the distance the body swings either to the right or the left of the rest position.

The regular back and forth motion of the mass in 1.12 describes what is known as a *harmonic oscillation*.

For $r > 0$, the trajectory $z(t)$ goes to zero as $t \to \infty$, and so the amplitude of the oscillations decrease over time. Also, the motion is no longer periodic when $r > 0$.

When damping is substantial, in the sense that $r^2/4 > \omega^2$, then the roots λ_i are real and no oscillations occur at all. Motion damps to zero in this case without any back and forth swings (Exercise 1.5.1).

The solution 1.11 for complex roots can be written more compactly as

$$z(t) = A \sin(\omega t + \phi), \tag{1.13}$$

for some ϕ. In fact by equating 1.11 and 1.13 and taking note of the trigonometric identity

$$\sin(\omega t + \phi) = \sin \omega t \cos \phi + \sin \phi \cos \omega t, \tag{1.14}$$

it follows that

$$c_1 = A \sin \phi \qquad \text{and} \qquad c_2 = A \cos \phi.$$

Therefore $c_1^2 + c_2^2 = A^2$ and $\tan\phi = c_1/c_2$. These relations determine the *amplitude* A and the so called *phase angle* ϕ completely. Depending on the context either 1.11 or 1.13 may be a more revealing way of expressing the solution.

Note that the constants c_1 and c_2 in 1.11 or A and ϕ in 1.13 are determined by specifying the initial values of x and $\dot{x}$. Although the choice $x(0) = x_0$ and $\dot{x}(0) = 0$ resulted in expression 1.12, the initial conditions could have been something else.

The upshot of this discussion is that the mass–spring system behaves as expected. With no resistance ($r = 0$) the mass oscillates in a regular back and forth manner about the rest position once it has been disturbed. With positive resistance (damping) the back and forth swings eventually die down and motion ultimately ceases.

We can now visualize a more interesting situation in which the mass and spring is held vertically in one of our slightly shaky hands. The fixed support beam is replaced by a support that itself moves up and down. For simplicity, we model the hand motion by a simple harmonic oscillation, which imparts a force on the mass–spring system given by $B \sin \omega_0 t$, where ω_0 is the frequency of hand motion and B is the amplitude (namely, the magnitude of either the up or down excursion). This motion results in an additional compression or stretching of the spring. By a suitable change of variables, the gravitational effect can be incorporated as in 1.7, and so the composite system has the equation

$$\ddot{z} + r\dot{z} + \omega^2 z = B \sin \omega_0 t, \tag{1.15}$$

which simply extends Equation 1.7 by including the additional external force. Equation 1.15 describes a *forced oscillator*. The solution of 1.15 and its interpretation are left to the exercises.

1.3 Stable Equilibria, I

The mathematical models considered up to now are special cases of

$$\ddot{z} + r\dot{z} + \omega^2 z = g(t), \tag{1.16}$$

where $g(t)$ is some function of time. It is expedient to rewrite 1.16 as a system of first-order equations. Most model formulations in this book occur naturally as a first-order system, whereas the second-order equations in this chapter are a direct result of Newton's law. In any event, it is easy to convert 1.16 into the desired form. Simply introduce two new variables x_1 and x_2 by letting $x_1 = z$ and $x_2 = \dot{z}$. It then follows immediately that

$$\begin{aligned} \dot{x}_1 &= x_2 \\ \dot{x}_2 &= -\omega^2 x_1 - r x_2 + g(t) \end{aligned} \tag{1.17}$$

or, in vector–matrix notation,

$$\dot{\mathbf{x}} = A\mathbf{x} + \mathbf{b}, \tag{1.18}$$

where

$$\mathbf{x} = \begin{pmatrix} x_1 \\ x_2 \end{pmatrix}, \quad \dot{\mathbf{x}} = \begin{pmatrix} \dot{x}_1 \\ \dot{x}_2 \end{pmatrix}, \quad \mathbf{b} = \begin{pmatrix} 0 \\ g(t) \end{pmatrix},$$

and

$$A = \begin{pmatrix} 0 & 1 \\ -\omega^2 & -r \end{pmatrix}.$$

As a generalization of 1.17 consider a pair of nonlinear first-order equations

$$\begin{aligned} \dot{x}_1 &= f_1(x_1, x_2) \\ \dot{x}_2 &= f_2(x_1, x_2) \end{aligned} \tag{1.19}$$

in which f_i are given smooth functions of $\mathbf{x}$ on some open subset U of $\mathbf{R}^2$. In vector notation, 1.19 is expressed as

$$\dot{\mathbf{x}} = \mathbf{f}(\mathbf{x}),$$

with

$$\mathbf{f}(\mathbf{x}) = \begin{pmatrix} f_1(\mathbf{x}) \\ f_2(\mathbf{x}) \end{pmatrix} = \begin{pmatrix} f_1(x_1, x_2) \\ f_2(x_1, x_2) \end{pmatrix}.$$

The vector $\mathbf{f}$ defines a *vector field* on U and a solution to 1.19 is a smooth curve in the plane whose tangent at $\mathbf{x}$ is $\mathbf{f}(\mathbf{x})$. The path of this curve is called a *trajectory* or, more commonly, an *orbit*. As t varies, the values of $\mathbf{x}(t)$ define the *states* of the *dynamical system* described by Equations 1.19 and U is called the *state space*.

Any orbit along which $\dot{\mathbf{x}}$ is identically zero is a fixed point of the motion and is referred to as an *equilibrium state*. In the second-order model 1.16, the equilibrium state of the corresponding first-order system 1.17 describes a situation in which there is no motion. This is due to the fact that since $x_1 = z$ and $x_2 = \dot{z}$, then $\dot{\mathbf{x}} = 0$ implies that velocity and acceleration are both zero.

Example 1.4 Consider the model of a vertical mass–spring system that led to the equation

$$\ddot{y} + r\dot{y} + \omega^2 y = -g.$$

That this is a special case of 1.16 is clear if we identify y with z and treat $g(t)$ as the constant $-g$. The equation is then equivalent to

$$\begin{aligned} \dot{x}_1 &= x_2 \\ \dot{x}_2 &= -\omega^2 x_1 - r x_2 - g. \end{aligned}$$

The single equilibrium state occurs at

$$\begin{aligned} x_1 &= -\frac{g}{\omega^2} \\ x_2 &= 0. \end{aligned} \tag{1.20}$$

Recall that the vertical position of rest was found earlier to have the value $y_0 = -g/\omega^2$, where $\dot{y} = 0$. It is entirely consistent that this rest point coincides with the equilibrium 1.20 since it designates a position of no motion in which all forces are in balance.

In connection with any equilibrium state $\bar{\mathbf{x}}$ of $\dot{\mathbf{x}} = \mathbf{f}(\mathbf{x})$ is the question of stability. This is an idea that is pervasive in this text and that will be introduced gradually over the next several chapters. The central notion is this. If one disturbs a physical system that is initially at rest, does it gradually return back to rest or does it wander away? More precisely, if one jolts the system to a new initial value, will the orbit tend to $\bar{\mathbf{x}}$ as $t \to \infty$, or will it at least remain in the vicinity of $\bar{\mathbf{x}}$? If it does, we say that $\bar{\mathbf{x}}$ is stable. Otherwise, it is unstable. In more mathematical terms, suppose that for every open neighborhood Ω of $\bar{\mathbf{x}}$ in U the orbit remains in Ω for all $t \geq 0$ whenever it starts close enough to $\bar{\mathbf{x}}$. Then $\bar{\mathbf{x}}$ is *stable*. Otherwise, it is *unstable*. In addition, if there is some open Ω_0 containing $\bar{\mathbf{x}}$ for which $\mathbf{x}(t) \to \bar{\mathbf{x}}$ as $t \to \infty$ whenever the orbit starts in Ω_0, then $\bar{\mathbf{x}}$ is

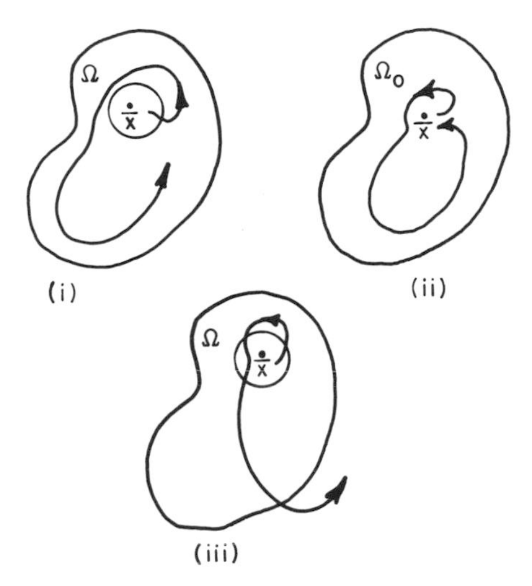

Figure 1.3 Examples of (i) stable, (ii) asymptotically stable, and (iii) unstable equilibria. In (iii) there is no neighborhood of $\bar{\mathbf{x}}$ that guarantees that an orbit remains within Ω.

called *asymptotically stable* (abbreviated a.s.). The largest possible Ω_0 that satisfies this property is called the *basin of attraction* of $\bar{\mathbf{x}}$. Figure 1.3 illustrates the possibilities. In the event that Ω_0 is the entire state space U, $\bar{\mathbf{x}}$ is said to be globally a.s. Thus, the equilibrium of the system in Example 1.4 is globally a.s. because $z(t)$ and $\dot{z}(t)$ both tend to zero as $t \to \infty$ for any initial displacement $z(0)$ and $\dot{z}(0)$. This much follows from the discussion in Section 1.2 since $r > 0$ is a damping factor which eventually forces all motion to cease.

It is too much to expect that, in general, equations of the form 1.19 can be solved explicitly. One must be prepared to use surrogate methods to understand the behavior of the orbits. The importance of the stability idea is that in a number of cases it allows us to deduce the wandering of the orbits in the long term without having to solve the equations themselves. For a stable equilibrium, we know that the orbit is likely to be found near this state of rest if the initial perturbation is mild enough. This information is often of great value. In Chapter Seven, for example, we study dynamic systems in which all orbits are expected to be at or near one of their stable equilibria after a sufficient lapse of time. Unstable equilibria represent states that one is unlikely to observe physically in such systems.

1.4 What Comes Out Is What Goes In

Up to now the model equations have been determined by the principle that the acceleration of a body is proportional to the forces that act on it. If there is no motion, these forces must balance to zero.

There is another useful principle that is also frequently used to derive differential equation models. It states that the rate at which a substance increases or decreases within some closed environment must be due solely to the rate at which matter enters from the outside minus the rate at which it leaves. This expresses the principle of *conservation of mass*. It is implicit in numerous models and is especially fundamental in the discussion of Chapter Five. By a closed environment is meant a region in which matter is neither created or destroyed or, to put it another way, in which there are no additional sources or sinks of the substance. If there are, then these need to be accounted for.

Let us look at some examples, each of which satisfies the relation:

$$\text{rate of change of a substance} = \text{input rate} - \text{output rate}.$$

Example 1.5 Water flows into a tank at a constant rate (volume per unit time). Water evaporates from the tank at a rate which we assume is proportional to $v^{2/3}$, where v is the current volume in the tank. In order to formulate a differential equation for v, we apply the conservation of mass idea. The rate of change of water volume in the tank is $\dot{v}$. It must be equal to the rate at which it enters minus the rate of exit (due, in this case, to evaporation). The input rate is a constant $k_1 > 0$ and the output rate is proportional to $v^{2/3}$. Hence

$$\dot{v} = k_1 - k_2 v^{2/3}, \tag{1.21}$$

with k_2 positive. The negative sign indicates a decrease in volume per unit time.

Note that 1.21 has an equilibrium state determined by $\dot{v} = 0$. It is given by $\bar{v} = (k_1/k_2)^{3/2}$. The decision as to whether it is stable or not can be inferred quite easily by looking at the sign of the derivative $\dot{v}$. If v is less than $\bar{v}$, then a glance at 1.21 shows that $\dot{v}$ is positive whereas its sign is negative when $v > \bar{v}$. This means that $v(t)$ increases (or decreases) toward the equilibrium state $\bar{v}$ depending on whether it is initially below (or above) the value $\bar{v}$. This shows that the rest point is globally a.s. It is important to observe that we arrived at this conclusion without explicitly solving the equation. A similar argument can be applied to many of the first-order nonlinear equations that are encountered later (as, for example, in Exercise 1.5.7). An extension of this idea, the isocline method, is taken up in Chapter Three.

Example 1.6 An organic pollutant enters a lake at a constant rate. Bacterial action metabolizes (decomposes) the pollutant at a rate proportional to its mass. In doing so, the dissolved oxygen in the lake waters is used up at the same rate that the pollutant decomposes. However, oxygen in the air re-enters the lake through surface-to-air contact (this is called re-aeration) at a rate proportional to the difference between the maximum dissolved oxygen level that the lake can support and its current actual value.

To find the dissolved oxygen level x_2 in the lake at any time, observe that its rate of change depends on an input which is proportional to $x_m - x_2$, where x_m is the maximum (saturation) level of oxygen in the water, and an output which is proportional to the mass x_1 of pollutant:

$$\dot{x}_2 = -kx_1 + k_1(x_m - x_2), \tag{1.22}$$

with k and k_1 positive. At the same time, the rate of change of x_1 depends on a constant input rate $\sigma > 0$ and a decay rate which is linear in x_1 itself:

$$\dot{x}_1 = \sigma - kx_1 \tag{1.23}$$

By solving for x_1 in 1.23 and inserting this into 1.22, it is apparent that 1.22 becomes a linear nonhomogeneous first-order equation in x_2 and hence solvable. We forgo the details since this model is encountered again in a more general setting in Chapter Five. Observe, however, that an equilibrium state occurs at $x_1 = \sigma/k$ and $x_2 = -\sigma/k_1 + x_m$, where the oxygen content of the water is less than optimum. Is it stable? The methods of the next chapter permit us to answer this (Exercise 2.5.7).

1.5 Exercises

1.5.1 Solve Equation 1.7 when $r/2 < \omega$, subject to the initial conditions $z(0) = 0$ and $\dot{z}(0) = 0$. Show that as $r \to 0$ one recovers the solution 1.12. When $r/2 > \omega$, establish that the mass returns to zero as $t \to \infty$ without swinging past the origin.

1.5.2 A raindrop falls from a cloud and is subject to a constant gravitational force. Air resistance provides viscous damping, but there is no restoring force, of course. Solve the differential equation that models this situation and interpret the solution.

1.5.3 Solve Equation 1.15. Since $r > 0$, we know from 1.10 that the homogeneous solution tends to zero as $t \to \infty$. Therefore, one expects that ultimately the solution depends entirely on the forcing function $B \sin \omega_0 t$. That is, the solution should eventually oscillate periodically with period ω_0. Taking a cue from the representation 1.13, try a particular solution in the form

$$z_p(t) = A \sin(\omega_0 t + \phi).$$

1.5.4 Interpret the solution obtained in the previous exercise. In particular, consider what happens when $\omega = \omega_0$ (this condition is called *resonance*). Also, consider the case of r being very small compared to ω. As the hand moves up and down, the mass on the spring either moves up and down synchronously or asynchronously with the hand, depending on whether ω is greater or less than ω_0. Show this.

1.5.5 The text discusses the stability of fixed points. However, one can also study the stability of some given orbit $\tilde{\mathbf{x}}(t)$ by letting $\mathbf{z}(t) = \mathbf{x}(t) - \tilde{\mathbf{x}}(t)$. Show that $\mathbf{z}$ satisfies a vector equation of the form $\dot{\mathbf{z}} = \mathbf{g}(\mathbf{z})$ for a suitable function $\mathbf{g}$ and that the stability of $\tilde{\mathbf{x}}(t)$ is equivalent to the stability of the fixed point $\mathbf{z} = 0$.

1.5.6 A spherical raindrop evaporates at a rate that is proportional to its surface area. Find an expression for the volume v of the drop as a function of time by solving a suitable differential equation.
Hint: If r is the raindrop radius, then v is proportional to r^3 while surface area is proportional to r^2.

1.5.7 Water enters a cylindrical tank at a constant rate (volume per unit time). The tank has a hole in the bottom and so water flows out through it. Let h be the depth of the water in the tank. It is known that the rate at which water flows through the hole (distance per unit time) is proportional to $\sqrt{h}$. Suppose the tank has a constant cross-sectional area A and that the hole has area A_h. Find a differential equation for h. Determine the equilibrium state and discuss whether it is stable or not (see Example 1.5 in the text).

1.5.8 Using separation of variables solve Equation 1.21 in the text and verify the a.s. of the equilibrium state.
Hint: Rewrite it as $\dot{x} = c(1 - x^{2/3})$ for some constant c and let $x^2 = y^6$.

Chapter **Two**

Stable and Unstable Motion, I

2.1 The Pendulum

An historically important nonlinear model is that of the pendulum in a plane. It will be the center of our discussion in this chapter.

The pendulum consists of a small bob of mass m attached to a slim, rigid, and massless rod of length l, which is itself connected to a pivot P. The bob can swing in a circular arc about the pivot. The straight down position finds the bob at rest. If it is displaced by an angle from the vertical and then let go, it is expected to swing in a circular arc in the plane back and forth about its rest position which is designated to be $\theta = 0$. Another rest position occurs in the straight up position at $\theta = \pi$, but experience tells us that even the slightest disturbance will make the bob move away from this position. We expect it to be unstable there.

When the pendulum is displaced from rest, the only restoring force is that of gravity, which tends to pull it back downward. Since motion is confined to lie along the circular arc in the plane of radius l about P, we are only interested in that component of gravitational force that is tangent to the arc (Figure 2.1). One can also allow for viscous damping and this too is taken to be tangential to the arc.

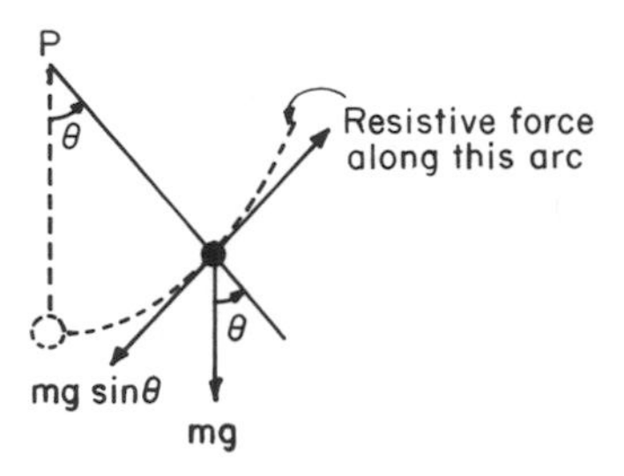

Figure 2.1 Force diagram for the pendulum. The restoring force along the arc due to gravity is $mg \sin \theta$.

Arc length s is defined by $s = l\theta$ with θ increasing in the counter-clockwise direction. Since l is constant, this means that $\dot{s} = l\dot{\theta}$ and $\ddot{s} = l\ddot{\theta}$. Acceleration of the bob along the arc is therefore $\ddot{s} = l\ddot{\theta}$. The resistive force is assumed to be proportional to the velocity along the arc. The resistive force is therefore $k_1\dot{s} = k_1 l\dot{\theta}$ for some constant k_1. Newton's law then states that

$$ml\ddot{\theta} = -k_1 l\dot{\theta} - mg \sin \theta. \tag{2.1}$$

The negative signs in 2.1 occur because the restoring force is in the negative θ direction while the damping force is opposite to the direction of motion. If $\theta > 0$, so that movement is downward along the arc, damping should be upward.

Letting $r = k_1/m$, 2.1 can be written as

$$\ddot{\theta} + r\dot{\theta} + \frac{g}{l} \sin \theta = 0. \tag{2.2}$$

This equation is nonlinear and cannot be solved explicitly. One can always integrate numerically on a computer, of course, for any given set of initial conditions on θ and $\dot{\theta}$. This generates numbers that may or may not be useful depending on what we are trying to learn about the movement of the pendulum. Our interest is in understanding something about where its mass will be found in the long run rather than on the specifics of how it wanders about in the short term. For example, will the displaced bob eventually return to its equilibrium state? This is a question of stability, and it can be resolved without solving Equation 2.2. This requires the use of surrogate methods to replace the unavailability of an explicit solution. The emphasis on indirect solution techniques to give insight into the qualitative behavior of the orbits will prove to be a key methodological weapon in the remainder of this book.

We begin by rewriting 2.2 as a system of two first-order equations following the prescription in Section 1.3. Let $x_1 = \theta$ and $x_2 = \dot{\theta}$. Then 2.2 is equivalent to

$$\begin{aligned} \dot{x}_1 &= x_2 \\ \dot{x}_2 &= -rx_2 - \frac{g}{l} \sin x_1. \end{aligned} \tag{2.3}$$

The equilibria of 2.3 occur where the right hand sides are zero. This requires that $x_2 = 0$ and $x_1 = \pm n\pi$ for $n = 0, 1, 2, \ldots$. Since $\sin x_1$ is periodic, the only truly distinct equilibrium states occur at $x_1 = 0, \pi$. In terms of the original variables, one therefore obtains essentially two fixed points of motion defined by

$$\begin{pmatrix} \theta \\ \dot{\theta} \end{pmatrix} = \begin{pmatrix} 0 \\ 0 \end{pmatrix} \quad \text{and} \quad \begin{pmatrix} \theta \\ \dot{\theta} \end{pmatrix} = \begin{pmatrix} \pi \\ 0 \end{pmatrix}.$$

Intuitively, we expect the first of these to be stable, the other unstable. In order to verify this mathematically, a short digression is necessary.

2.2 When Is a Linear System Stable?

Suppose we are given a system of two linear first-order homogeneous differential equations

$$\begin{aligned} \dot{u}_1 &= a_{11}u_1 + a_{12}u_2 \\ \dot{u}_2 &= a_{21}u_1 + a_{22}u_2 \end{aligned}$$

in which the a_{ij} are constants. In vector notation, the equations are written as

$$\dot{\mathbf{u}} = A\mathbf{u}, \tag{2.4}$$

where A is the matrix

$$\begin{pmatrix} a_{11} & a_{12} \\ a_{21} & a_{22} \end{pmatrix}$$

and

$$\mathbf{u}(t) = \begin{pmatrix} u_1(t) \\ u_2(t) \end{pmatrix}.$$

The components of the vector $\mathbf{u}$ describe the orbits of 2.4 in the u_1, u_2 plane.

Assume that A is nonsingular and that its eigenvalues λ_1 and λ_2 are distinct. This is the typical or generic situation and a reasonable modeling assumption. We defer any explanation of what is meant by generic until Chapter Three.

The only equilibrium of 2.4 occurs at the origin since A is nonsingular. The solution to 2.4 is given by

$$\mathbf{u}(t) = \alpha_1 e^{\lambda_1 t} \mathbf{c}_1 + \alpha_2 e^{\lambda_2 t} \mathbf{c}_2 \tag{2.5}$$

in which $\mathbf{c}_1, \mathbf{c}_2$ are the linearly independent eigenvectors of A corresponding to the eigenvalues λ_1 and λ_2 (see Appendix). Suppose, first, that the λ_i are real. From 2.5, it is apparent that if the λ_i are negative, then $\mathbf{u}(t) \to \mathbf{0}$ as $t \to \infty$ no matter where the starting point $\mathbf{u}(0)$ is located. Therefore, the origin is globally asymptotically stable (a.s.).

Similarly, if the λ_i are positive, then one is forced to conclude that $\mathbf{u}(t) \to \infty$ as t increases. In this case, the origin repels all orbits and is unstable.

When the eigenvalues are complex with $\lambda_i = \sigma \pm iq$ then 2.5 becomes

$$\mathbf{u}(t) = e^{\sigma t}(\zeta_1 \cos qt + \zeta_2 \sin qt) \tag{2.6}$$

in which the ζ_i are suitable vectors (the reader is again referred to the Appendix for details). A glance at 2.6 now reveals that if $\sigma = \operatorname{Re} \lambda_i < 0$, then $\mathbf{u}(t)$ spirals down to zero from any starting point, and so the origin is globally a.s.; whereas $\sigma > 0$ implies that the origin is repelling and unstable.

In the event that $\sigma = 0$, the solution is evidently bounded. This is called the *neutrally stable* case. The orbits are in fact closed just as in the solution of the undamped mass–spring system of Chapter One. For each different $\mathbf{u}(0)$, there is a corresponding closed orbit (called a *cycle*) passing through this point, and so the plane is filled with concentric cycles. None of these cycles intersect each other. The reason is that *all differential equations in this book, linear and nonlinear, are assumed to be smooth enough to guarantee uniqueness of solutions*. Therefore, at most one orbit can pass through any given point in the plane. To put it another way, orbits cannot cross each other. This uniqueness idea is invoked numerous times in subsequent chapters, and it will play a significant role in our modeling work.

Suppose, finally, that only one of the eigenvalues has a negative real part. Then the λ_i are necessarily real (why?) and we find, say, that

$$\lambda_1 < 0 < \lambda_2.$$

From relation 2.5, it is apparent that if $\mathbf{u}(0)$ is a multiple of $\mathbf{c}_1$, then $\mathbf{u}(t) \to \mathbf{0}$ as $t \to \infty$, whereas $\mathbf{u}(t) \to \infty$ if $\mathbf{u}(0)$ is initially a multiple of $\mathbf{c}_2$. Hence, the origin is unstable. A linear combination of these two motions leads to saddle-shaped orbits that appear to approach the origin and then move away. The equilibrium state is called a *saddle*. Figure 2.2 illustrates the saddle and other typical cases. These facts are summarized in

Lemma 2.1 *The zero equilibrium of $\dot{\mathbf{u}} = A\mathbf{u}$ is globally asymptotically stable if and only if the real parts of the eigenvalues of A are negative and unstable otherwise.*

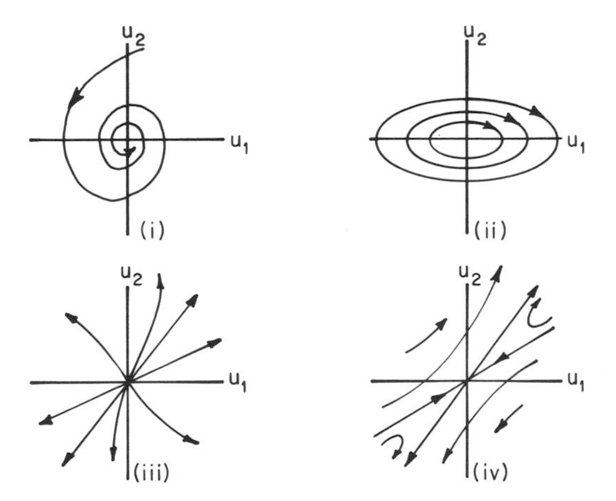

Figure 2.2 Typical orbits of the linear system 2.4: (i) complex eigenvalues with $\sigma < 0$ (asymptotic stability); (ii) complex eigenvalues with $\sigma = 0$ (neutral stability); (iii) real eigenvalues, both negative (unstable); (iv) real eigenvalues, one positive and the other negative (unstable saddle).

Instead of computing the eigenvalues of A there is a shortcut. It is not difficult to show the following equivalence (Exercise 2.5.1):

Lemma 2.2 *The eigenvalues* λ_i *of the* 2×2 *matrix* A *satisfy* $\operatorname{Re} \lambda_i < 0$ *if and only if* $\lambda_1 \lambda_2 = \operatorname{Det} A > 0$ *and* $\lambda_1 + \lambda_2 = \operatorname{Trace} A < 0$. *They are pure imaginary if and only if the trace is zero. Moreover,* $\lambda_1 < 0 < \lambda_2$ *(or* $\lambda_2 < 0 < \lambda_1$*) if and only if* $\operatorname{Det} A < 0$.

Example 2.1 Let us reconsider the equation of a damped mass–spring system:

$$\ddot{z} + r\dot{z} + \omega^2 z = 0. \tag{2.7}$$

Let $u_1 = z$ and $u_2 = \dot{z}$. Then 2.7 can be rewritten, by our now familiar procedure, as a first-order system of the form 2.4 in which

$$A = \begin{pmatrix} 0 & 1 \\ -\omega^2 & -r \end{pmatrix}. \tag{2.8}$$

The eigenvalues of A are easily computed to be

$$\lambda_i = -\frac{r}{2} \pm \sqrt{\frac{r^2}{4} - \omega^2},$$

and so $\operatorname{Re} \lambda_i < 0$. Thus, the rest position $\mathbf{u} \equiv \mathbf{0}$ or, equivalently, $z = \dot{z} = 0$, is asymptotically stable. The same conclusion is reached via Lemma 2.2 since $\operatorname{Det} A = \omega^2 > 0$ and $\operatorname{Trace} A = -r < 0$.

Without damping (the situation in which $r = 0$) the eigenvalues have zero real parts, and we can only conclude neutral stability from Lemma 2.1. Indeed this is the case of the harmonic oscillator that was shown in Chapter One to have a bounded periodic motion.

2.3 When Is a Nonlinear System Stable?

Let us return to Equation 2.2 or, rather, its expression as a first-order system 2.3. This is a special case of a general autonomous system of the form

$$\begin{aligned} \dot{x}_1 &= f_1(x_1, x_2) \\ \dot{x}_2 &= f_2(x_1, x_2), \end{aligned} \tag{2.9}$$

where f_i are smooth functions. Suppose $\bar{\mathbf{x}} = \begin{pmatrix} \bar{x}_1 \\ \bar{x}_2 \end{pmatrix}$ is a unique equilibrium state of 2.9, and let $u_i = x_i - \bar{x}_i$ represent the deviations of x_i from this state. From the well-known Taylor's Theorem for functions of two variables, it is known that

$$f_i(x_1, x_2) = f_i(\bar{x}_1, \bar{x}_2) + \frac{\partial f_i}{\partial x_1}(\bar{x}_1, \bar{x}_2)u_1 + \frac{\partial f_i}{\partial x_2}(\bar{x}_1, \bar{x}_2)u_2 + g_i(u_1, u_2), \tag{2.10}$$

for $i = 1, 2$. The g_i are higher order terms, which go to zero faster than u_1 and u_2. Specifically, if $\|\mathbf{u}\| = \sqrt{u_1^2 + u_2^2}$ is the length of the vector $\mathbf{u}$, then

$$\frac{g_i(u_1, u_2)}{\|\mathbf{u}\|} \to 0 \quad \text{as} \quad \|\mathbf{u}\| \to 0. \tag{2.11}$$

A proof of this result is to be found in texts on advance calculus and need not be given here.

Since $\bar{\mathbf{x}}$ is an equilibrium state, $f_i(\bar{x}_1, \bar{x}_2) = 0$ for $i = 1, 2$, and one also readily sees that $\dot{u}_i \equiv \dot{x}_i$. It follows that 2.10 can be expressed in vector notation as

$$\dot{\mathbf{u}} = A\mathbf{u} + \mathbf{g}(\mathbf{u}), \tag{2.12}$$

where

$$\mathbf{u} = \begin{pmatrix} u_1 \\ u_2 \end{pmatrix}, \qquad \mathbf{g}(\mathbf{u}) = \begin{pmatrix} g_1(u_1, u_2) \\ g_1(u_1, u_2) \end{pmatrix},$$

and

$$A = \begin{pmatrix} \frac{\partial f_1}{\partial x_1}(\bar{x}_1, \bar{x}_2) & \frac{\partial f_1}{\partial x_2}(\bar{x}_1, \bar{x}_2) \\ \frac{\partial f_2}{\partial x_1}(\bar{x}_1, \bar{x}_2) & \frac{\partial f_2}{\partial x_2}(\bar{x}_1, \bar{x}_2) \end{pmatrix} = \left. \begin{pmatrix} \frac{\partial f_1}{\partial x_1} & \frac{\partial f_1}{\partial x_2} \\ \frac{\partial f_2}{\partial x_1} & \frac{\partial f_2}{\partial x_2} \end{pmatrix} \right|_{\mathbf{x}=\bar{\mathbf{x}}}. \tag{2.13}$$

The matrix A is the *Jacobian matrix* of $\mathbf{f}$ at $\bar{\mathbf{x}}$.

When $\mathbf{x}(t)$ is close to $\bar{\mathbf{x}}$, $\mathbf{g}(\mathbf{u})$ is "small". In this case, 2.12 is approximated by the linear system

$$\dot{\mathbf{u}} = A\mathbf{u}, \tag{2.14}$$

which is called the *linearized system* corresponding to 2.9.

One can now imagine the following situation. Because of some disturbance, the equilibrium state $\bar{\mathbf{x}}$ is perturbed to a new initial value $\mathbf{x}(0)$. We would like to know if $\bar{\mathbf{x}}$ is asymptotically stable or, to put it another way, whether $\mathbf{x}(t) \to \bar{\mathbf{x}}$ as $t \to \infty$. But $\mathbf{u}(t) = \mathbf{x}(t) - \bar{\mathbf{x}}$, and so our question reduces to that of knowing if $\mathbf{u}(t) \to \mathbf{0}$. Since 2.14 approximates 2.9, we are now asking whether the zero state of $\dot{\mathbf{u}} = A\mathbf{u}$ is asymptotically stable. *One must be careful.* The approximation was conditioned on $\mathbf{x}$ being close to $\bar{\mathbf{x}}$. So we are talking here of the *local stability* of $\bar{\mathbf{x}}$ in the sense of some initial jolt that leaves $\mathbf{x}(t)$ not too far from its place of rest. Just how local is "local" is to be examined in Section 3.2. However, for the moment we can reason intuitively that if the linearized system has a globally a.s. equilibrium, then at least locally the same must be true for the equilibrium $\bar{\mathbf{x}}$ of the nonlinear system 2.9. As a result of Lemma 2.1 in the previous section, one can tentatively conclude that $\bar{\mathbf{x}}$ is locally a.s. whenever the matrix A of the linearized system has eigenvalues with negative real part.

If $\bar{\mathbf{u}} = \mathbf{0}$ is neutrally stable, then no conclusion can be drawn about $\bar{\mathbf{x}}$, as an example will later show. Also when $\bar{\mathbf{u}} = \mathbf{0}$ is unstable, it is again not possible to decide on the long term behavior of $\mathbf{x}(t)$ since the distance between $\mathbf{x}$ and $\bar{\mathbf{x}}$ increases as $\mathbf{u}(t) \to \infty$, and so the approximation 2.14 loses its validity.

Confirmation of our intuition is provided in Section 3.2, where it is shown that if the eigenvalues of the linearized system have nonzero real parts (in which case $\bar{\mathbf{x}}$ is called a *hyperbolic equilibrium*), the global a.s. of the origin in 2.14 implies the local a.s. of $\bar{\mathbf{x}}$ in 2.9.

For a hyperbolic equilibrium, there is an even stronger result which states that the orbits of the nonlinear system in a neighborhood of $\bar{\mathbf{x}}$ are qualitatively similar to those of the linear system about the origin. More precisely, the orbits about $\bar{\mathbf{x}}$ can be continuously deformed in some neighborhood of $\bar{\mathbf{x}}$ into the corresponding orbits of the linear system near the origin. This is a fairly deep theorem in analysis, known as the *Hartman–Grobman theorem*. Moreover, if $\bar{\mathbf{x}}$ is a saddle point, there are a pair of orbits of 2.9 that approach (move away from) $\bar{\mathbf{x}}$ and whose

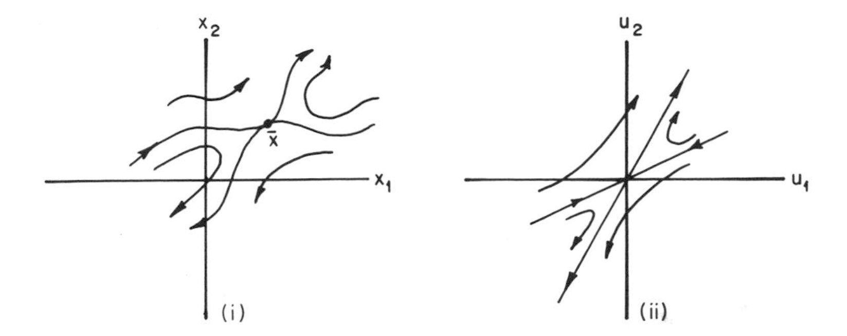

Figure 2.3 Orbit behavior near a saddle point for (i) the nonlinear system 2.9 and (ii) its linearization.

tangent at $\bar{\mathbf{x}}$ is an eigenvector of A corresponding to the negative (positive) eigenvalue (Figure 2.3). In modern parlance, these orbits are known, respectively, as the *stable and unstable manifolds* of 2.9 at $\bar{\mathbf{x}}$. The stable manifold is also called a *separatrix*. We will not attempt to prove these assertions. Instead the results will be used to guide our intuition when trying to interpret the local behavior of orbits (the reader is directed to the References for details on where proofs may be found).

Example 2.2 Let us apply these insights to the pendulum Equations 2.3. Here $f_1(x_1, x_2) = x_2$ and $f_2(x_1, x_2) = -rx_2 - (g/l)\sin x_1$. Therefore

$$A = \left.\begin{pmatrix} \dfrac{\partial f_1}{\partial x_1} & \dfrac{\partial f_1}{\partial x_2} \\ \dfrac{\partial f_2}{\partial x_1} & \dfrac{\partial f_2}{\partial x_2} \end{pmatrix}\right|_{\mathbf{x}=\bar{\mathbf{x}}} = \left.\begin{pmatrix} 0 & 1 \\ -\dfrac{g}{l}\cos x_1 & -r \end{pmatrix}\right|_{\mathbf{x}=\bar{\mathbf{x}}}$$

Consider first the equilibrium state $\bar{\mathbf{x}} = \begin{pmatrix} 0 \\ 0 \end{pmatrix}$. Then

$$A = \begin{pmatrix} 0 & 1 \\ -\dfrac{g}{l} & -r \end{pmatrix},$$

and so Det $A = g/l > 0$ and Trace $A = -r < 0$. From Lemma 2.2 one concludes that $\bar{\mathbf{x}}$ is locally a.s. However, if $\bar{\mathbf{x}} = \begin{pmatrix} \pi \\ 0 \end{pmatrix}$, then

$$A = \begin{pmatrix} 0 & 1 \\ \dfrac{g}{l} & -r \end{pmatrix},$$

so that Det $A = -g/l < 0$ and Trace $A = -r < 0$. In this case, Lemma 2.2 tells us that the linearized system has an unstable saddle equilibrium.

It is conceivable that even if $\bar{\mathbf{x}}$ is locally not a.s., it still remains bounded and hence stable. The local linear analysis given above is not strong enough to reveal the behavior of $\mathbf{x}(t)$ as $t \to \infty$ as we have already mentioned. Nevertheless in certain cases, it is possible to show that the orbit does indeed follow some well defined stable path. An example of this occurs below in Section 2.4 and, in a different way, in Section 6.1.

2.4 The Phase Plane

Another approach to understanding the qualitative behavior of nonlinear models may be considered in terms of a single second-order scalar differential equation of the form

$$\ddot{p} + f(p, \dot{p}) = 0. \tag{2.15}$$

An example of 2.15 is the pendulum Equation 2.2 in which $p = \theta$.

To begin with, it is expedient to suppose that f is independent of $\dot{p}$. This gives

$$\ddot{p} + f(p) = 0. \tag{2.16}$$

An illustration is found by letting $r = 0$ in 2.2.

Motivated by the fact that up to now we have been discussing mechanical systems in motion, it is useful to think of 2.15 or 2.16 as describing the motion of some body along a coordinate direction p. Inextricably bound to this idea is the concept of energy. To see what this means, multiply 2.16 by $\dot{p}$ and integrate to obtain

$$\tfrac{1}{2}\dot{p}^2 + U(p) = \text{constant}, \tag{2.17}$$

where

$$U(p) = \int_0^p f(s)\,ds. \tag{2.18}$$

Students of physics recognize $\frac{1}{2}\dot{p}^2$ as the *kinetic energy* of the motion, whereas $U(p)$ is called the *potential energy*. Then 2.17 expresses the fact that *total energy* E (kinetic plus potential) must be constant along the orbits of 2.16. Note that the derivative of E is $\dot{E} = \dot{p}(\ddot{p} + f(p)) = 0$, which is another way of showing that E is constant, a fact known as the principle of *conservation of energy*. It relates the variables p and $\dot{p}$, and so it helps us to deduce the orbits of 2.16 in a number of cases. Observe that if $\ddot{p} + f(p) = 0$ is written as an equivalent system of first-order equations by letting $x_1 = p$ and $x_2 = \dot{p}$ then the p, $\dot{p}$ plane is identical to the x_1, x_2 plane. The orbits of 2.16 are said to belong to the p, $\dot{p}$ *phase plane*.

The best way to see how to use the notion of energy conservation is through examples.

Example 2.3 The simplest illustration is provided by the familiar undamped mass–spring model along the z axis. Now, however, the variable z is denoted by p:

$$\ddot{p} + \omega^2 p = 0. \tag{2.19}$$

Here

$$U(p) = \omega^2 \int_0^p s\,ds = \frac{\omega^2 p^2}{2}.$$

Therefore, total energy is

$$E = \frac{\dot{p}^2 + \omega^2 p^2}{2}. \tag{2.20}$$

Since E is constant, 2.20 is the equation of an ellipse in the p, $\dot{p}$ plane. As E varies, we get a family of concentric ellipses.

Let us look at this orbit picture in a slightly different way. Plot the potential energy $U(p)$ against p. From 2.17, the energy E obeys the inequality

$$E \geq U(p).$$

Therefore, $U(p) = E$ if and only if $\dot{p} = 0$. Moreover, $\dot{p}$ increases in magnitude as $U(p)$ decreases from E, reaching its maximum when $U(p)$ is at its lowest point at $p = 0$. Thereafter, U again rises up towards the value E as p itself increases. As this happens, $\dot{p}$ decreases in magnitude until it reaches zero. By projecting the values of $U(p)$ onto the p, $\dot{p}$ phase plane, it is easy to deduce that the orbits must have the shape of concentric ellipses, one ellipse for each value of E (Figure 2.4). The direction of the arrows in Figure 2.4 can be understood by the fact that $\dot{p}$ is positive as p increases ($\dot{p} < 0$ for decreasing p).

What the figure shows is that p oscillates back and forth in a regular manner, which corresponds, of course, to the known solution of a harmonic oscillator. One can also read off from the figure that the velocity $\dot{p}$ is maximum as the body passes the zero position $p = 0$. If

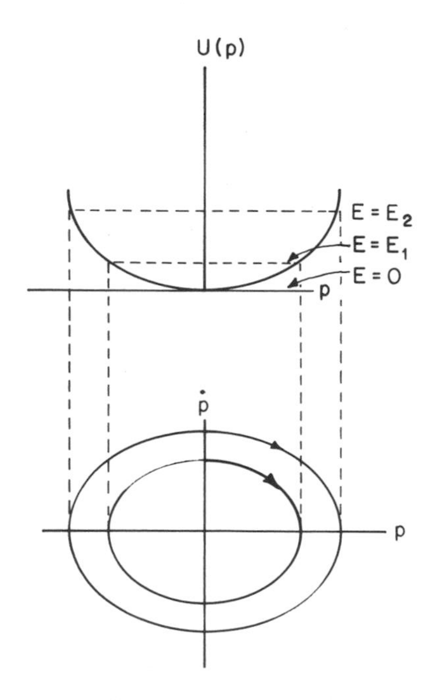

Figure 2.4 Phase portrait of Equation 2.19 (harmonic oscillator). The orbits are shown for energy levels $0 < E_1 < E_2$. When $E = 0$, the orbit is a point, the equilibrium state.

2.19 had been written as a first-order system by letting $x_1 = p$ and $x_2 = \dot{p}$, then it is immediately noticed that the sole equilibrium occurs at the origin of the phase plane where $p = \dot{p} = 0$. This corresponds to the energy level $E = 0$.

E is constant, and so $E = \frac{1}{2}[\omega^2 p^2(0) + \dot{p}(0)]$. E therefore changes as $p(0)$ and $\dot{p}(0)$ vary. Thus, the more one stretches (or compresses) the spring and the greater the initial velocity is in magnitude, the larger is E. The corresponding orbit will also be larger. Physically, of course, we cannot push or pull too much without violating the assumption of an ideal spring (after all, it may break!) but mathematically the $p, \dot{p}$ phase plane describes all possible motions with increasingly larger elliptical orbits as E increases. Observe that the origin in the phase plane possesses neutral stability.

Example 2.4 Now consider the undamped pendulum in which θ is written as p. The equation of motion is

$$\ddot{p} + \frac{g}{l} \sin p = 0.$$

Total energy is

$$E = \frac{\dot{p}^2}{2} + \frac{g}{l} \int_0^p \sin \zeta \, d\zeta = \frac{\dot{p}^2}{2} + \frac{g}{l}(1 - \cos p). \tag{2.21}$$

As in the previous example, we plot the potential energy $U(p) = (g/l)(1 - \cos p)$ versus p and, directly below it, the $p, \dot{p}$ phase plane corresponding to different energy levels. $E = 0$ forces $\dot{p}$ to be zero and p to take on one of the values for which $\cos p = 1$, namely $p = \pm 2\pi n$ for $n = 0, 1, 2, \ldots$ (Figure 2.5). These positions of zero potential energy correspond to the known stable equilibrium states of the pendulum in the $\theta, \dot{\theta}$ plane (see Section 2.1).

As E increases beyond zero, the orbits cycle about the stable equilibria, provided total energy E remains less than the critical value E_2 shown in the figure. The orbits have a shape roughly similar to concentric ellipses. In fact, since 2.21 implies $U(p) \leq E$, $\dot{p}$ is necessarily zero if $U(p) = E < E_2$. However, as $U(p)$ decreases away from E, the magnitude of $\dot{p}$ increases reaching its maximum where $U(p)$ is a minimum. Thereafter, $|\dot{p}|$ gets smaller as $U(p)$ begins to rise back up to the total energy level E. Since $\dot{p}$ is positive (negative) as p increases

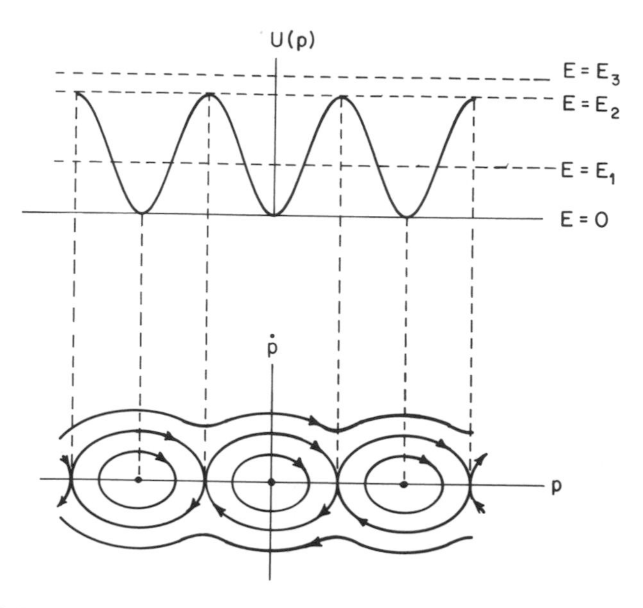

Figure 2.5 Phase portrait of the undamped pendulum. The orbits are shown for energy levels $0 < E_1 < E_2 < E_3$. The minimum potential energy occurs at $E = 0$.

(decreases), the direction of the orbits is clockwise as shown by the arrows. Such motion reflects the back and forth swings of the pendulum about its stable equilibrium.

Because $U(p) = E_2$ at $p = \pm\pi n$, the relation $U(p) \leq E_2$ together with the knowledge that total energy is E_2 combine to show that $\dot{p} = 0$ at these points. As we know from our earlier discussion in Section 2.1, such points are the unstable equilibria of the pendulum in the $\theta, \dot{\theta}$ plane. By the same reasoning given above, the orbits corresponding to total energy E_2 connect the unstable points to each other. But now it is essential to recognize that since these points are also solutions to the pendulum differential equation, the orbits can only approach the unstable equilibria asymptotically without actually reaching them in any finite time. Otherwise, the solutions would eventually cross and this would violate the principle of uniqueness.

It must therefore take an infinite amount of time to reach (or move away from) these points as p increases (decreases). This implies that motion along such orbits slows down as it gets closer to or moves away from the equilibria. In effect, it says that if the undamped pendulum bob is slightly displaced from the unstable upright position in which $\theta = \pi$,

$\dot{\theta} = 0$, it would require an infinite time to swing past the bottom and back up to the top.

If total energy exceeds E_2, say $E = E_3$, then $U(p)$ never equals E_3 and so $\dot{p}$ is nonzero. However, it is still true that as $U(p)$ increases (decreases) the magnitude of $\dot{p}$ will correspondingly get smaller (larger). This is reflected in the orbits shown in the figure. In terms of the pendulum, this says that if we initially push the bob hard enough (choose $\dot{p}(0)$ large enough), then the undamped pendulum continues to swing around past the top position and back around again and again.

The shape of the orbits at energy level E_2 indicates that the unstable equilibria are saddles linked to each other by stable and unstable manifolds. This fits in nicely with our earlier observation that the linearized system has indeed an unstable saddle equilibrium (Example 2.2).

It is now fairly apparent that without actually solving the pendulum equation a great deal of qualitative information can be read off from the phase plane portrait. One final remark about this example is that although this portrait repeats itself *ad infinitum* along the p axis it suffices to consider a single interval of length 2π to understand what is going on. A pictorial representation that is sometimes convenient can be obtained by taking a vertical slice of the phase plane of width 2π and then rolling it into a cylinder. If one edge of the slice is at a saddle point, then all unstable equilibria merge into this one. This is known as the cylindrical phase space representation of the pendulum, but we do not show it here (see, however, the references to Chapter Two).

Example 2.5 Let us consider now a physically absurd but mathematically instructive mass–spring system with a nonlinear restoring force that expands the spring as we compress it! The equation of motion is

$$\ddot{p} + \omega^2 p^2 = 0. \tag{2.22}$$

Total energy is given by

$$E = \frac{\dot{p}^2}{2} + \omega^2 \int_0^p s^2\, ds = \frac{\dot{p}^2}{2} + \frac{\omega^2 p^3}{3}.$$

As before, we plot the potential energy $U(p) = \frac{1}{3}\omega^2 p^3$ and, below it, the p, $\dot{p}$ phase plane (Figure 2.6). In this example, it makes sense to have negative energy levels, as shown. Also none of the orbits are closed since

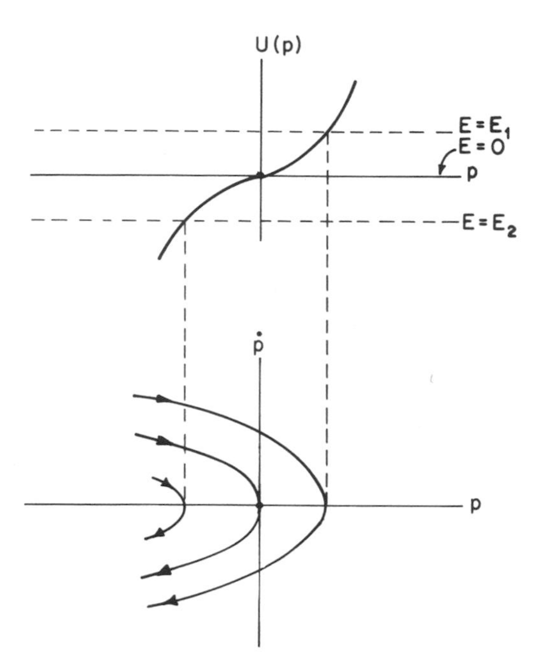

Figure 2.6 Phase portrait of Equation 2.22. The orbits are unbounded since energy can take on any value.

$U(p)$ equals the total E only at a single point p, for each E. At this point, $\dot{p}$ is of course zero. As $U(p)$ decreases away from E, $\dot{p}$ increases in magnitude indefinitely. By writing 2.22 as a system of first-order equations, it is clear that the sole equilibrium occurs at $p = \dot{p} = 0$. Hence, the orbit corresponding to the energy level zero can only approach the origin asymptotically but not cross it. The reason is uniqueness of orbits. Hence, this orbit slows down as it approaches the origin, and it requires an infinite amount of time to do so. All other orbits cross the p axis (where $\dot{p} = 0$) in finite time and continue past. The phase portrait indicates an unstable equilibrium. A linearized stability analysis (Exercise 2.5.2) verifies this.

Finally, note that since $p \to -\infty$ as $t \to \infty$ (except for $E = 0$), $\dot{p}$ eventually increases without bound in magnitude. This signifies that motion along the orbits is faster and faster as t increases.

In all our examples so far total energy has been constant. Dynamic systems of the form

$$\ddot{p} + f(p) = 0$$

are therefore called *conservative*. However, if f depends on $\dot{p}$ as well as p, then this need no longer be true. For instance, total energy in the pendulum equation in which damping is included may decrease. To verify this recall that the expression for E is

$$E = \frac{\dot{p}^2}{2} + \frac{g}{l}(1 - \cos p).$$

Therefore

$$\dot{E} = \dot{p}\left(\ddot{p} + \frac{g}{l}\sin p\right). \tag{2.23}$$

However, the damped pendulum equation (with θ replaced by p) is

$$\ddot{p} + r\dot{p} + \frac{g}{l}\sin p = 0 \tag{2.24}$$

(this is Equation 2.2). Substituting 2.24 into 2.23 gives

$$\dot{E} = -r\dot{p}^2$$

and so $\dot{E}$ is nonpositive, which shows that E never increases and may even decrease. A system in which energy is lost is called *dissipative*.

In the next chapter, the situation $\dot{E} \leq 0$ will be treated more carefully using the notion of a Liapunov function. It will be seen that the orbits of the damped pendulum equation tend to a rest point as a result of a decrease in total energy. The system runs itself down because of energy dissipation due to damping.

Let us close by giving a last example. Some familiarity with elementary electric circuit theory is required to understand how the present model comes about. However, in the later chapters the equation will arise in a different way without the use of circuit theory, and so if you are not familiar with this topic, there is no cause for alarm. Just concentrate on the equation itself.

Example 2.6 An electrical circuit consists of a resistor, an inductor, and a capacitor connected together by wires (Figure 2.7). A source of electrical energy supplies a voltage difference v_C between a_1 and a_2. This causes a current to flow. Now imagine that this voltage source is removed. What happens to the current flow? To answer this let us invoke the modeling assumption known as *Kirchhoff's voltage law*. It states that the voltage v_C is equal to the sum of voltage difference between a_1 and

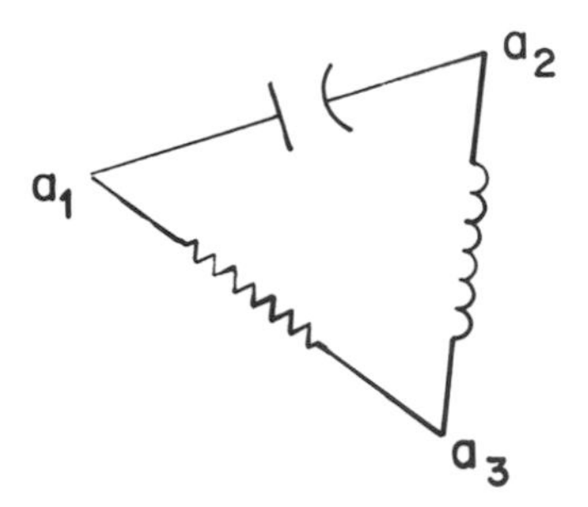

Figure 2.7 A simple circuit consisting of a capacitor, inductor, and a nonlinear resistor.

a_3, called v_L, and a_3 and a_2, called v_R:

$$v_C = v_L + v_R. \tag{2.25}$$

The capacitor, inductor, and resistor are physical elements in which voltage is related in a certain way to the current flowing in them. These relations somewhat idealize what is actually measured in a laboratory but are good approximations within a certain range of values for voltage and current. It is good modeling practice to choose postulates that are reasonably compatible with what is actually observed and, at the same time, permit a certain mathematical simplicity. This is similar to the way we treated the mass–spring system in Chapter One and the pendulum earlier in the present chapter, in which springs were considered to be ideal, gravitational attraction constant, and resistance a linear function of velocity. These are useful fictions which allow the modeler to penetrate more readily into the heart of the matter.

In any event, let us postulate the following relations between the voltages and the current u in the circuit:

$$\begin{aligned} v_L &= \dot{u} \\ v_C &= -\int_0^t u(s)\,ds \\ v_R &= \alpha\left(\tfrac{1}{3}u^3 - u\right), \qquad \alpha > 0. \end{aligned} \tag{2.26}$$

The last relation in 2.26 requires comment since the voltage across a resistor is usually expressed as a linear function of u (Ohm's law). However, in vacuum tube or semiconductor circuits, a nonlinear relationship appears to be more appropriate. Since we are looking to

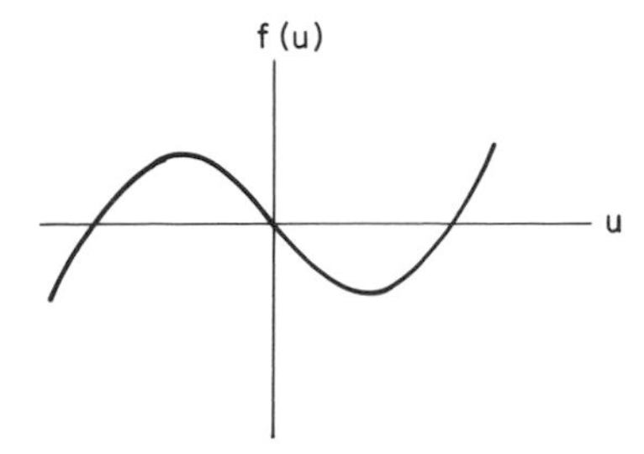

Figure 2.8 The function $f(u) = \alpha(\frac{1}{3}u^3 - u)$, $\alpha > 0$.

formulate a nonlinear model, let us assume that in fact our circuit is of this type. The plot of $f(u) = \alpha(\frac{1}{3}u^3 - u)$, where α is a positive constant, is given in Figure 2.8. It approximates the way the resistance is actually observed as a function of current in certain circuits. Inserting 2.26 into 2.25 gives us the equation

$$\dot{u} + \alpha\left(\frac{u^3}{3} - u\right) + \int_0^t u(s)\,ds = 0. \tag{2.27}$$

By differentiating once, 2.27 becomes

$$\ddot{u} + u + \alpha\dot{u}(u^2 - 1) = 0, \tag{2.28}$$

which is known as the *Van der Pol equation*. Thinking back in terms of the spring–mass system, we could interpret 2.28 as a spring with linear restoring force, as before, but with a resistance term that behaves as a damping force when $u^2 > 1$ and, most curiously, in just the opposite manner when $u^2 < 1$. This says that for small movements of the spring motion is encouraged rather than impeded, but that as soon as this same motion gives rise to large excursions from the equilibrium state then genuine damping takes place. This seems difficult to accept for the spring model, but it is a reasonable occurrence in certain circuits having vacuum tubes and semiconductors. If $u^2 > 1$, the system behaves dissipatively, but $u^2 < 1$ suggests that energy is being pumped in. The implications of this give and take kind of situation are to be explored later in Chapter Six.

Note that if we define v to be v_C, then 2.26 shows that $\dot{v} = -u$. Combining this with 2.27 gives a first-order system equivalent to the Van der Pol equation:

$$\dot{u} = v - \alpha\left(\frac{u^3}{3} - u\right) \tag{2.29}$$
$$\dot{v} = -u.$$

The only equilibrium of 2.29 occurs at $u = v = 0$. It is left to Exercise 2.5.5 to study its local stability.

The phase plane analysis of dissipative systems will not be attempted until we have introduced the geometrical device of isoclines in the next chapter.

2.5 Exercises

2.5.1 Prove Lemma 2.2 by showing that if A is the 2×2 matrix

$$A = \begin{pmatrix} a_{11} & a_{12} \\ a_{21} & a_{22} \end{pmatrix},$$

then its eigenvalues λ_1 and λ_2 satisfy Trace $A = \lambda_1 + \lambda_2$ and Det $A = \lambda_1\lambda_2$. Moreover, Re $\lambda_i < 0$ for $i = 1, 2$ if and only if Trace $A < 0$ and Det $A > 0$. The determinant is negative if and only if the λ_i are real with one negative and the other positive.

2.5.2 Show that the equation $\ddot{p} + \omega^2 p^2 = 0$ has a locally unstable equilibrium by using linearization.

2.5.3 Consider the model

$$\ddot{p} + \omega^2 p + \alpha p^3 = 0$$

in which α can be positive or negative. Determine the equilibrium and by linearization do a local stability analysis in both cases.

2.5.4 Do a phase plane analysis of the equation in the previous exercise. Compare to the results on stability obtained there by linearization.

2.5.5 Determine whether the equilibrium of the Van der Pol Equation 2.28 is locally stable or not.

2.5.6 Consider the undamped pendulum, whose equation is

$$\ddot{p} + \frac{g}{l}\sin p = 0.$$

Total energy is given by relation 2.21:

$$E = \frac{\dot{p}^2}{2} + U(p),$$

where $U(p) = (g/l)(1 - \cos p) \leq E$. Therefore

$$\dot{p} = \pm\sqrt{2[E - U(p)]}\,. \tag{2.30}$$

Since $U(p) \leq E$, we are taking the square root of a positive quantity in 2.30. Now let us compute the total time T it takes the orbit to move along the separatrix in Figure 2.5 from a given initial position $p_0 > -\pi$ to a final position $p_1 < \pi$. Since $\dot{p} \geq 0$ along such a path, the sign of the square root in 2.30 is taken to be positive. From 2.30 we obtain, by the usual change of variable formula in integral calculus,

$$T = \int_0^T dt = \int_{p_0}^{p_1} \frac{dp}{\dot{p}} = \int_{p_0}^{p_1} \frac{dp}{\sqrt{2[E - U(p)]}}.$$

Now E is constant along the orbit and so it suffices to compute it at a single point, say $\begin{pmatrix} p_1 \\ \dot{p}_1 \end{pmatrix}$. Then

$$T = \int_{p_0}^{p_1} \frac{dp}{\sqrt{\frac{2g}{l}(\cos p - \cos p_1) + \dot{p}_1^2}}. \tag{2.31}$$

Reason why T is unbounded as $p_1 \to \pi$ in 2.31. This is an alternate argument to show that the pendulum requires an infinite amount of time to go around once from its upright unstable state back to where it started from (Example 2.4).

2.5.7 Discuss the stability of the equilibrium in Example 1.6 of Chapter One.

2.5.8 Do a phase plane analysis of a mass–spring system in the case of Coulomb friction (Equation 1.4). By manipulating the relation for total energy show that the orbits are two different sets of ellipses depending on the sign of $\dot{\mathbf{x}}$. Deduce from the phase plane diagram that motion ceases in finite time. (This problem will be reconsidered in Section 6.1.)

Chapter **Three**

Stable and Unstable Motion, II

3.1 Liapunov Functions

In Section 2.3, it was seen how linearization about an equilibrium state $\bar{\mathbf{x}}$ can sometimes yield local stability information. There is an alternate approach that can be quite useful but depends on being able to find a suitable function in a neighborhood of $\bar{\mathbf{x}}$.

We begin with the nonlinear system 2.9, which is defined for all $\mathbf{x}$ in some open set U in $\mathbf{R}^2$. For convenience it is re-stated here:

$$\begin{aligned} \dot{x}_1 &= f_1(x_1, x_2) \\ \dot{x}_2 &= f_2(x_1, x_2). \end{aligned} \tag{3.1}$$

Suppose $\bar{\mathbf{x}} = \begin{pmatrix} \bar{x}_1 \\ \bar{x}_2 \end{pmatrix}$ is an isolated equilibrium of 3.1 in U and that there is a smooth scalar function V defined on U such that $V(\bar{\mathbf{x}}) = 0$ and $V(\mathbf{x}) > 0$ for $\mathbf{x} \neq \bar{\mathbf{x}}$. If l is any nonnegative constant, then the set of points $\mathbf{x}$ for which $V(\mathbf{x}) = l$ is called a *level curve*. Visually a level curve is the projection onto U of a slice of V having elevation l. Since V is single valued, the level curves corresponding to different l do not intersect, and one sees them as wrapped about $\bar{\mathbf{x}}$ (Figure 3.1).

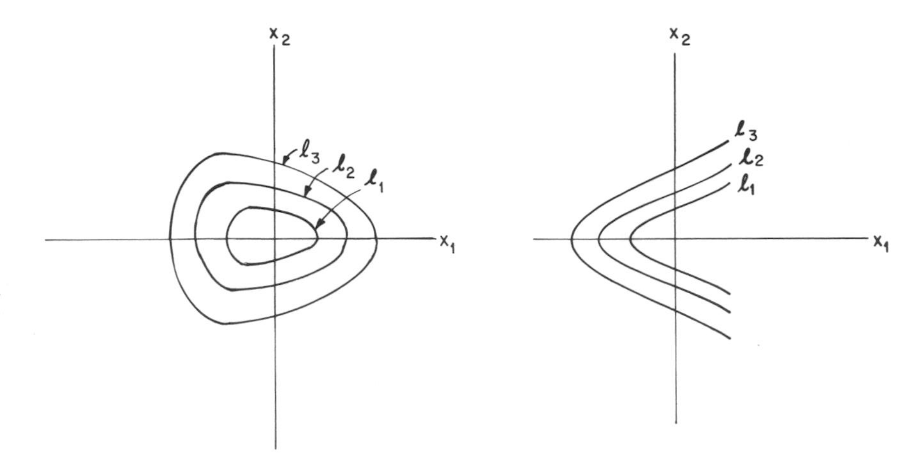

Figure 3.1 Level curves in $\mathbf{R}^2$ of two nonnegative functions V that are zero only at $\bar{\mathbf{x}} = \mathbf{0}$. Here $l_1 < l_2 < l_3$.

For any open set Ω about $\bar{\mathbf{x}}$, consider the subset Ω_0 of Ω consisting of all points in a sufficiently small neighborhood of $\bar{\mathbf{x}}$ for which $V(\mathbf{x}) < l$. Suppose orbits never cross a level curve in an outward direction. Then if an orbit begins in Ω_0, it remains trapped within this set and hence stays in Ω. Therefore $\bar{\mathbf{x}}$ is stable. This is the intuitive content of Theorem 3.1 below.

The argument seems quite simple, but unfortunately it rests on being able to find an appropriate function V. In some cases, this can be done and that is why the yet to be proven theorem is useful. A simple example illustrates the essential idea:

Example 3.1 Consider the harmonic oscillator equation

$$\ddot{p} + \omega^2 p = 0.$$

Letting $x_1 = p$ and $x_2 = \dot{p}$, this can be written as a first-order system

$$\begin{aligned} \dot{x}_1 &= x_2 \\ \dot{x}_2 &= -\omega^2 x_1. \end{aligned} \tag{3.2}$$

The equilibrium occurs at $\bar{\mathbf{x}} = \mathbf{0}$. A choice of V is simple. Let V be the total energy E (recall the definition in Section 2.4). Then

$$V(x_1, x_2) = \frac{x_2^2}{2} + \frac{\omega^2 x_1^2}{2}.$$

This function is defined for all $\mathbf{x}$ in $\mathbf{R}^2$ and is clearly positive except at $\bar{\mathbf{x}} = \mathbf{0}$. The level curves are concentric ellipses in the x_1, x_2 plane (the $p, \dot{p}$ phase plane) which get smaller as the total energy level decreases. Along a solution curve $\mathbf{x}(t)$, the *derivative of V with respect to time* is defined by the chain rule of differentiation as

$$\dot{V} = \frac{dV(\mathbf{x}(t))}{dt} = \frac{\partial V(\mathbf{x})}{\partial x_1}\dot{x}_1 + \frac{\partial V(\mathbf{x})}{\partial x_2}\dot{x}_2 = x_2\dot{x}_2 + \omega^2 x_1 \dot{x}_1,$$

which is zero because of 3.2. This shows that any orbit of 3.2 is confined to a level curve of V and that V is non-increasing (actually constant) along any of these solution curves. This means that if an orbit begins inside a level curve of V, where total energy is l, it cannot escape outside to higher energy levels. Having entered inside a given elliptical contour, it remains inside for all time and so the origin is stable.

It is time for a formal definition. Let $\Omega \subset U$ be an open set in which $\bar{\mathbf{x}}$ is an isolated equilibrium of 3.1. Suppose we can find a smooth nonnegative function V on Ω that is zero only at $\bar{\mathbf{x}}$. Then V is called a *Liapunov function* if its derivative with respect to t along solution curves of 3.1 is nonpositive for all t:

$$\frac{dV(\mathbf{x}(t))}{dt} \le 0.$$

There may be many such functions on Ω. The multiplication of any given V by a positive scalar is again a Liapunov function, for example.

In establishing stability of $\bar{\mathbf{x}}$, it suffices to show that any orbit which begins in a suitably small subset of Ω will remain in Ω (recall the discussion in Section 1.3). This implies that the orbit is also trivially confined to any open subset of U which contains Ω.

Before giving the next theorem, a few more preliminaries are necessary. First, we recall that if $\bar{\mathbf{x}}$ is an equilibrium state, it represents a solution orbit $\mathbf{x}(t) \equiv \bar{\mathbf{x}}$ for all t. Therefore, unless a given orbit of 3.1 is already at $\bar{\mathbf{x}}$, it cannot reach $\bar{\mathbf{x}}$ in finite time without violating the uniqueness of solutions. Second, define an *invariant set* in U to be one for which any orbit that begins in the set remains within it for all future time. An equilibrium state, for example, is invariant. If R is a set in U containing no invariant subsets, then whenever an orbit enters R it must leave again after a finite amount of time.

Theorem 3.1 *Let* $\bar{\mathbf{x}}$ *be an isolated equilibrium of* 3.1 *in some open set* Ω *in* U. *Let* V *be a Liapunov function on* Ω. *Then* $\bar{\mathbf{x}}$ *is stable. Suppose* R *is the subset of* Ω *in which* $\dot{V} = 0$. *If* R *contains no invariant subsets other than* $\bar{\mathbf{x}}$ *itself, then* $\bar{\mathbf{x}}$ *is asymptotically stable.*

Proof Denote by $\|\mathbf{x} - \bar{\mathbf{x}}\|$ the distance in $\mathbf{R}^2$ between $\mathbf{x}$ and $\bar{\mathbf{x}}$. Let B be the neighborhood of $\bar{\mathbf{x}}$ given by

$$\|\mathbf{x} - \bar{\mathbf{x}}\| < \delta.$$

If $\delta > 0$ is small enough, then B lies within Ω. Let γ be the minimum of V on the boundary of B. Then $\gamma > 0$ since V is zero only at $\bar{\mathbf{x}}$. Denote by Ω_0 the set of all points in B for which $V(\mathbf{x}) < \gamma$. Since V is a continuous function that shrinks to zero as $\mathbf{x}$ gets close to $\bar{\mathbf{x}}$, it follows that Ω_0 is not empty.

Pick any solution of 3.1 in which the initial point $\mathbf{x}(0)$ lies in Ω_0. Then the orbit never crosses the boundary of B since $\dot{V} \leq 0$ means that V is non-increasing along the orbit. Thus, $\mathbf{x}(t)$ lies within Ω_0 for all $t \geq 0$, which is stability.

If, for some $t > 0$, $\mathbf{x}(t)$ lies outside of the closed set R, then

$$\frac{dV(\mathbf{x}(t))}{dt} = \lim_{s \to 0} \frac{V(\mathbf{x}(t+s)) - V(\mathbf{x}(t))}{s} < 0.$$

It follows that $V(\mathbf{x}(t))$ is locally decreasing in the direction of the orbit at time t or, to put it another way, the orbit itself is moving locally inward across the level curve defined by V at that point. So, outside of R, V is strictly decreasing. Moreover, the orbit cannot be stalled indefinitely within R. Sooner or later it leaves. So where does it end up? It cannot remain in R (unless it begins at $\bar{\mathbf{x}}$) nor can it come to rest outside of R, and so it has no choice to move asymptotically toward $\bar{\mathbf{x}}$ where V attains its lower bound of zero.

One must be a bit more careful. Since V is non-increasing and bounded below, we know that $V(\mathbf{x}(t)) \to \eta \geq 0$ as $t \to \infty$. However, it is necessary to be convinced that η is actually zero and that in fact this value occurs at $\mathbf{x} = \bar{\mathbf{x}}$. We regard the argument given in the previous paragraph to be sufficiently obvious that it is not necessary to fuss over details. A complete proof can be found in several of the references quoted.

It is an immediate corollary of the above theorem that if $\dot{V} < 0$ everywhere on Ω except at $\bar{\mathbf{x}}$, then $\bar{\mathbf{x}}$ is asymptotically stable.

Also worthy of note is that $\bar{\mathbf{x}}$ is unstable if the function V satisfies

$$\frac{dV(\mathbf{x}(t))}{dt} > 0$$

is some neighborhood of $\bar{\mathbf{x}}$. The reason, of course, is that by a reversal of the previously used argument, we now can infer that an orbit cuts level curves outwardly in a direction of increasing V. The orbit therefore moves away from $\bar{\mathbf{x}}$.

Example 3.2 Let us reconsider the pendulum equation with damping:

$$\ddot{\theta} + r\dot{\theta} + \frac{g}{l} \sin \theta = 0. \tag{3.3}$$

Written as a first-order system 3.3 becomes

$$\begin{aligned} \dot{x}_1 &= x_2 \\ \dot{x}_2 &= -rx_2 - \frac{g}{l} \sin x_1. \end{aligned} \tag{3.4}$$

The point $\bar{\mathbf{x}} = \mathbf{0}$ (that is, $\theta = \dot{\theta} = 0$) is an isolated equilibrium in the region Ω defined by $|x_1| < \pi$. As a Liapunov function, it is again expedient to choose total energy as in Example 3.1. Therefore, we let

$$V(\mathbf{x}) = \frac{x_2^2}{2} + \frac{g}{l}(1 - \cos x_1) \tag{3.5}$$

(see Section 2.4). It is apparent from 3.5 that $V(\mathbf{x}) > 0$ on Ω except at the origin where it is zero. From 3.4, a simple computation shows that

$$\dot{V} = x_2\dot{x}_2 + \frac{g}{l}\dot{x}_1 \sin x_1 = x_2\left(\dot{x}_2 + \frac{g}{l} \sin x_1\right) = -rx_2^2. \tag{3.6}$$

Therefore, $\dot{V} = 0$ on the set R in Ω consisting of all $\mathbf{x}$ for which $x_2 = 0$ (namely the x_1 axis in the x_1, x_2 plane). But when $x_2 = 0$, 3.4 becomes

$$\begin{aligned} \dot{x}_1 &= 0 \\ \dot{x}_2 &= -\frac{g}{l} \sin x_1, \end{aligned}$$

which shows that $\dot{x}_2$ is nonzero on R except at $x_1 = 0$. Having entered the x_1 axis, the orbit does not remain on it since $\dot{x}_2 \neq 0$ obliges the motion to be transversal to R. Hence R contains no invariant subsets except the origin. By Theorem 3.1 the equilibrium is asymptotically stable.

It is important to understand that the asymptotic stability is only local in nature. The theorem guarantees that the orbit tends to the origin only if it begins in some suitable subset Ω_0 of Ω. Points that lie outside of Ω_0 could eventually leave Ω and not tend to zero. Physically, this corresponds to the fact that if the pendulum is given a very hard push, it will go around the bottom position several times before coming to rest. So it bypasses the equilibrium $\bar{\mathbf{x}} = \mathbf{0}$ and ends up instead at one of the rest points $\begin{pmatrix} 2\pi k \\ 0 \end{pmatrix}$ for $k \neq 0$. Thus, even though the initial displacement satisfies $|x_1| < \pi$, the final position can find x_1 outside of Ω.

Example 3.3 In Example 2.6 we introduced the Van der Pol equation, which can be written as a first-order system

$$\dot{x}_1 = -x_2$$

$$\dot{x}_2 = x_1 - \alpha\left(\frac{x_2^3}{3} - x_2\right).$$

(These are Equations 2.31 if $x_1 = v$ and $x_2 = u$.) Exercise 2.5.5 showed that the equilibrium $\bar{\mathbf{x}} = \mathbf{0}$ is unstable for $\alpha > 0$. What about $\alpha < 0$?

Define a Liapunov function by

$$V(\mathbf{x}) = \frac{x_1^2 + x_2^2}{2},$$

which is positive for all $\mathbf{x}$ except at the origin where it is zero. Moreover,

$$\dot{V} = -\alpha x_2^2\left(\frac{x_2^2}{3} - 1\right).$$

Since α is negative, the condition $\dot{V} \leq 0$ requires that $x_2^2 \leq 3$. This defines the set Ω on which V is Liapunov. We can also characterize the set Ω_0 in a fairly straightforward way as the disc of radius $\sqrt{3}$ in the x_1, x_2 plane. It is clear, in fact, that $\Omega_0 \subset \Omega$. Moreover, V increases in value as $\mathbf{x}$ moves radially outward so that an orbit that begins in Ω_0 remains there for all time.

The subset of Ω on which $\dot{V} = 0$ has only the origin as an invariant subset. This is shown in the same way as in the previous example. Hence the origin is a.s. and Ω_0 belongs to its domain of attraction.

Another choice for V, one that is suggested by analogy with total energy, is

$$V(\mathbf{x}) = \frac{x_2^2 + \dot{x}_2^2}{2}. \tag{3.7}$$

Exercise 3.4.2 asks for Ω and Ω_0 in this case.

Example 3.4 Let us return now to the conservative system

$$\ddot{p} + f(p) = 0.$$

Written as first-order equations, this becomes

$$\begin{aligned} \dot{x}_1 &= x_2 \\ \dot{x}_2 &= -f(x_1). \end{aligned} \tag{3.8}$$

Suppose $\bar{x}_1$ is a root of $f(x_1) = 0$ and that $\bar{\mathbf{x}} = \begin{pmatrix} \bar{x}_1 \\ 0 \end{pmatrix}$ is an isolated equilibrium of 3.8 in some region Ω. Consider the modified energy function

$$V(\mathbf{x}) = \frac{x_2^2}{2} + U(x_1) - U(\bar{x}_1), \tag{3.9}$$

where $U(x_1)$ is the potential function

$$U(x_1) = \int_0^{x_1} f(s)\, ds.$$

Suppose that $\bar{x}_1$ minimizes U on Ω. Then $V(\mathbf{x}) \geq 0$ on Ω with zero only at $\bar{\mathbf{x}}$. Moreover, along any orbit,

$$\frac{dV(\mathbf{x}(t))}{dt} = x_2\dot{x}_2 + \dot{x}_1 f(x_1) = x_2[\dot{x}_2 + f(x_1)] = 0$$

since $x_1 = p$ and $x_2 = \dot{p}$. By Theorem 3.1, $\bar{\mathbf{x}}$ is stable. This establishes the principle that any local equilibrium of a conservative system that minimizes the potential must be stable. In all the examples so far in which $U(x)$ was plotted, the minimum value of U indeed occurs at a stable point. Observe, incidently, that the level curves of V are actually the orbits of $\ddot{p} + f(p) = 0$.

Although the Liapunov approach tells us that orbits never bend outward, transversal to the level curves of V, the actual path can sometimes be inferred by a simple intuitive device known as the *isocline method*.

There are two sets of curves in the plane, called *isoclines*, obtained by taking $\dot{x}_1$ and $\dot{x}_2$ to be zero separately in 3.1. These are the level curves γ_1 and γ_2 given by $f_1(x_1, x_2) = 0$ and $f_2(x_1, x_2) = 0$. Their intersections are, of course, equilibrium points.

Along γ_i the rate of change of x_i is zero. Geometrically, this means that the orbit $\mathbf{x}(t)$ cuts the isocline γ_1 in a direction perpendicular to the x_1 axis while γ_2 is cut perpendicular to the x_2 axis. This observation often gives enough information to enable one to sketch the actual orbit. In later chapters, the isocline method will prove to be of considerable utility. For the moment an example will suffice.

Example 3.5 The mass–spring system with damping

$$\ddot{z} + r\dot{z} + \omega^2 z = 0 \tag{3.10}$$

has a first-order representation

$$\begin{aligned} \dot{x}_1 &= x_2 \\ \dot{x}_2 &= -\omega^2 x_1 - r x_2. \end{aligned} \tag{3.11}$$

Total energy in the undamped case in which $r = 0$ suggests a Liapunov function:

$$V(\mathbf{x}) = \frac{x_2^2 + \omega x_1^2}{2}.$$

The level curves of V are ellipses about the sole equilibrium $\bar{\mathbf{x}} = \mathbf{0}$. It is not difficult to show, using Theorem 3.1, that the origin is globally a.s. (Exercise 3.4.3).

The isoclines are obtained from 3.11. Curve γ_1 is the set where $x_2 = 0$ and γ_2 is defined by the straight line $x_2 = -(\omega^2/r)x_1$. The direction of

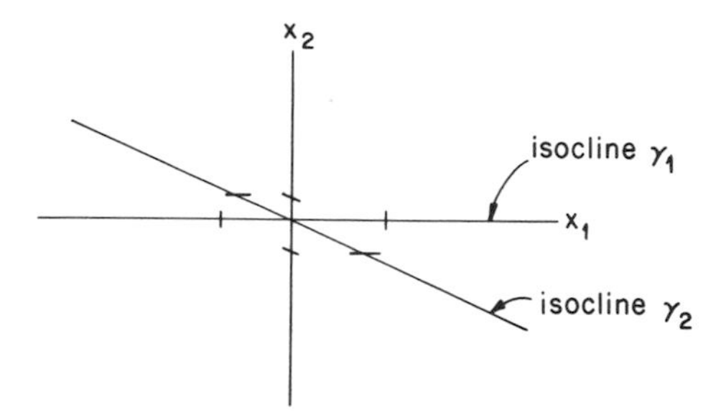

Figure 3.2 Orbit directions across the isoclines of 3.11.

any orbit across these curves must be perpendicular to one or the other axis. In addition, observe that $\dot{x}_2 = -rx_2$ when $x_1 = 0$ so that when the orbit cuts the x_2 axis, it does so with negative slope (Figure 3.2).

Connecting the small fragments of orbit shown in Figure 3.2 into a smooth trajectory suggests the picture given in Figure 3.3. The direction of the arrows is obtained by observing that in the first quadrant of the plane, where x_1 and x_2 are both positive, the rate of change of x_1 is positive while that of x_2 is negative. This much follows by simply looking at Equations 3.11. Therefore, the slope of the orbit is downward

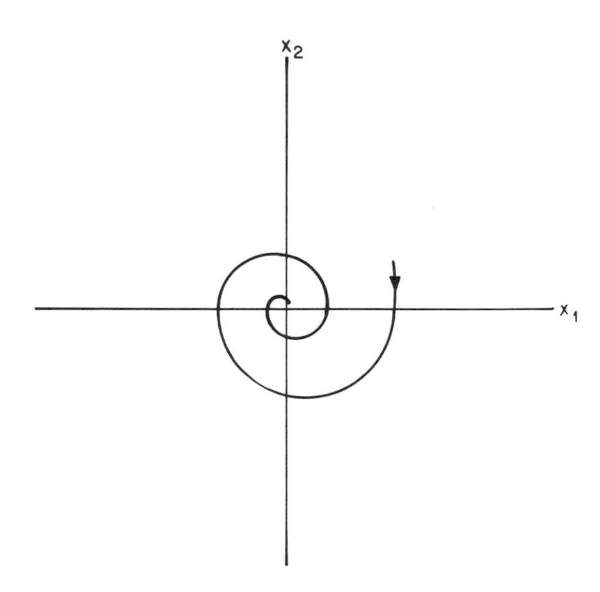

Figure 3.3 Phase portrait of 3.11 in which x_1 is position, and x_2 is velocity.

and to the right in this quadrant, meaning that movement is clockwise. Combining this picture with the knowledge that the origin is a.s. tells us that any orbit spirals down to zero.

3.2 Stable Equilibria, II

The discussion that follows is a bit more technical. Its primary intent is to validate the linearization procedure given in Section 2.3. The reader may wish to skip this section on a first reading.

We are given an isolated equilibrium $\bar{\mathbf{x}}$ of 3.1 in some region Ω. By letting $\mathbf{u} = \mathbf{x} - \bar{\mathbf{x}}$ an equivalent system is obtained that satisfies the vector system

$$\dot{\mathbf{u}} = A\mathbf{u} + \mathbf{g}(\mathbf{u}), \tag{3.12}$$

where $\mathbf{g}(\mathbf{u})$ are higher order terms which satisfy

$$\frac{g_i(\mathbf{u})}{\|\mathbf{u}\|} \to 0 \qquad \text{as } \|\mathbf{u}\| \to 0 \tag{3.13}$$

for each component g_i of the vector $\mathbf{g}$. (We suggest you refer to the discussion in Section 2.3 for the details in case it is not clear what this means.) The Jacobian matrix A is given by 2.13. Assume, for simplicity, that the eigenvalues λ_i of A are real, distinct, and negative. Lemma 2.1 guarantees that the zero equilibrium state of the linearized system

$$\dot{\mathbf{u}} = A\mathbf{u}$$

is globally a.s. What about the nonlinear system 3.12?

An answer is provided by constructing an appropriate Liapunov function. Let

$$V(\mathbf{u}) = \frac{u_1^2 + u_2^2}{2}.$$

Then, from 3.12, one obtains

$$\dot{V}(\mathbf{u}(t)) = u_1\dot{u}_1 + u_2\dot{u}_2 = \mathbf{u} \cdot A\mathbf{u} + \mathbf{u} \cdot \mathbf{g}(\mathbf{u}), \tag{3.14}$$

where the dot is the usual scalar or "dot" product of vectors in $\mathbf{R}^2$.

Let us assume that the linearly independent eigenvectors $\mathbf{v}_1$ and $\mathbf{v}_2$ corresponding to the distinct eigenvalues of A are orthonormal. When this is not true, the argument is lengthier but the result is the same. Every vector $\mathbf{u}$ can be written uniquely in terms of $\mathbf{v}_1, \mathbf{v}_2$:

$$\mathbf{u}(t) = c_1(t)\mathbf{v}_1 + c_2(t)\mathbf{v}_2. \tag{3.15}$$

An easy calculation shows that

$$A\mathbf{u} = \lambda_1 c_1 \mathbf{v}_1 + \lambda_2 c_2 \mathbf{v}_2$$

(since $A\mathbf{v}_i = \lambda_i \mathbf{v}_i$ for each i). Hence

$$\mathbf{u} \cdot A\mathbf{u} = \lambda_1 c_1^2 + \lambda_2 c_2^2 \leq \bar{\lambda}\left(c_1^2 + c_2^2\right)$$

in which $\bar{\lambda} = \max(\lambda_1, \lambda_2) < 0$. But one sees immediately from 3.15 that

$$c_1^2 + c_2^2 = \mathbf{u} \cdot \mathbf{u} \equiv \|\mathbf{u}\|^2.$$

From these results and the well-known Schwarz inequality for vectors, we therefore obtain from 3.14 that

$$\dot{V}(\mathbf{u}(t)) \leq \bar{\lambda}\|\mathbf{u}\|^2 + \|\mathbf{u}\| \cdot \|\mathbf{g}(\mathbf{u})\|. \tag{3.16}$$

However, the condition 3.13 implies if $\eta > 0$ is smaller than the magnitude of $\bar{\lambda}$ then

$$\|\mathbf{g}(\mathbf{u})\| < \eta\|\mathbf{u}\|$$

as soon as $\|\mathbf{u}\| < \delta$ for some suitably small $\delta > 0$. The inequality 3.16 then shows that if $\mathbf{u}$ lies in a sufficiently small neighborhood Ω_0 of the origin, then

$$\dot{V} \leq \|\mathbf{u}\|^2(\bar{\lambda} + \eta) < 0.$$

This verifies that V is a Liapunov function on Ω_0, and so the solutions 3.12 tend to zero as $t \to \infty$ by Theorem 3.1. This assures the a.s. of the nonlinear system 3.12 and therefore of 3.1. A similar proof goes through even if λ_i are complex. We give this as a theorem.

Theorem 3.2 *Let $\bar{\mathbf{x}}$ be an isolated equilibrium of $\dot{\mathbf{x}} = \mathbf{f}(\mathbf{x})$, where $\mathbf{f}$ is a smooth function in some subset of $\mathbf{R}^2$. If the eigenvalues of the linearized system have negative real parts, then $\bar{\mathbf{x}}$ is locally a.s.*

If $\operatorname{Re}\lambda_i = 0$, linearization offers no help in deciding stability. One sees this explicitly through an example. Let

$$\begin{aligned} \dot{x}_1 &= x_2 - \varepsilon x_1(x_1^2 + x_2^2) \\ \dot{x}_2 &= -x_1 - \varepsilon x_2(x_1^2 + x_2^2), \end{aligned} \tag{3.17}$$

where $\varepsilon > 0$. The only equilibrium is at the origin (Exercise 3.4.5), and the linearized system is

$$\begin{aligned} \dot{u}_1 &= u_2 \\ \dot{u}_2 &= -u_1. \end{aligned} \tag{3.18}$$

System 3.18 is simply the familiar undamped mass–spring equation in which $\omega^2 = 1$.

The A matrix of the linear system is

$$A = \begin{pmatrix} 0 & 1 \\ -1 & 0 \end{pmatrix},$$

and its eigenvalues are $\pm i$. Therefore, 3.18 has a neutrally stable equilibrium. Now if ε is quite small, then 3.17 is "close" to 3.18 and yet its stability properties are very different. To verify this, multiply the first equation of 3.17 by x_1 and the second equation by x_2. Then add the equations to get

$$x_1\dot{x}_1 + x_2\dot{x}_2 = -\varepsilon(x_1^2 + x_2^2)^2. \tag{3.19}$$

The left side of 3.19 is

$$\frac{1}{2}\frac{d}{dt}(x_1^2 + x_2^2) = \frac{1}{2}\frac{d}{dt}\|\mathbf{x}\|^2,$$

while the right side transforms to $-\varepsilon\|\mathbf{x}\|^4$. Therefore, 3.17 reduces to

$$\frac{d}{dt}\|\mathbf{x}\| = -\varepsilon\|\mathbf{x}\|^3. \tag{3.20}$$

This equation is separable and can be integrated. We find that

$$\frac{1}{\|\mathbf{x}(t)\|^2} = -2\varepsilon t + \alpha,$$

for some constant α. As $t \to \infty$, it follows that $\|\mathbf{x}(t)\|$ must go to zero, while if t approaches the negative value $-\alpha/2\varepsilon$, then $\|\mathbf{x}(t)\| \to \infty$. In one case the origin is asymptotically stable, and in the other it is unstable. And so even the slightest alteration in the form of 3.17 can radically alter the behavior of the corresponding orbit.

We have been discussing the stability of equilibria. There is another stability idea called *structural stability*. Loosely speaking structural stability is a property of a differential equation model whenever a small perturbation of the system equations leads to solution orbits that are qualitatively the same as before. Since models are often fairly crude descriptions of reality, it would be nice to know that a "roughly similar" model gives "roughly similar" orbits. This robustness would appear to be a reasonable requirement for any plausible model. For example a small change in the value of the coefficients of a differential equation should not significantly alter the nature of resulting solutions except perhaps to shift the equilibrium a bit. After all, such coefficients (like r and ω^2 in the damped spring–mass system) are found experimentally, and one should expect some error to creep in without altering drastically the kind of conclusions that can be drawn from the model.

The belief that good models are those that are structurally stable has been called the *stability dogma*. And yet, although it appears to be a reasonable request, it is often denied. The example given above shows, for instance, that even the most minute perturbation in the form of the equations

$$\begin{aligned} \dot{x}_1 &= x_2 \\ \dot{x}_2 &= -x_1 \end{aligned} \tag{3.21}$$

destroys the character of the orbits. Adding a very small damping term to 3.21 changes the nature of orbits from closed orbits about the origin to trajectories which spiral down to zero.

Equation 3.21 describes a mass–spring system, with frequency ω equal to unity, and we have assumed all along that it is a valid model. What is wrong? Well, perhaps one is asking too much. Equations 3.21 model a conservative system. Adding a damping term makes it dissipative. Moreover, the changes introduced in 3.17 by the addition of terms (with ε arbitrarily small) makes a linear system into a nonlinear one. It would be more reasonable to accept model changes that again lead to linear conservative systems. If the system was nonlinear to begin with, then perhaps only those perturbations that preserve the conservative nature of the model should be allowed.

In Chapter Four, a similar situation will arise when discussing a nonlinear ecological model that is again structurally unstable. Nonetheless, the model is instructive even if somewhat too simple. Its lack of robustness may be a blemish only if one demands too much from the equations.

In Chapters Seven and Eight, it will be seen that even systems that are for the most part robust can behave in a surprising manner as certain model parameters are varied.

While we are at it, let us dispose of another modeling assumption that has been used by us all along. Equation coefficients, as remarked above, are generally approximations to parameter values that are not precisely known. Eigenvalues of a linearized system can therefore vary if these parameters are slightly altered. Specifically, suppose one is given a linearized system

$$\dot{\mathbf{u}} = A\mathbf{u}$$

in which the entries of the Jacobian A can take on any value. A generic property of this system is one that holds most of the time. Technically, a statement about this class of linearized models is *generic* if it holds at least on a dense open subset of all such matrices. One would have to be very unlucky to find a matrix for which it is not true. In this context one can show, although we do not do so here, that it is generic for the set of square matrices A to have distinct eigenvalues. It is also generic that the equilibrium of a linear system is hyperbolic (the eigenvalues of A have nonzero real parts). As a corollary, nonsingular matrices are generic (Exercise 3.4.11). Generic properties simplify the analysis of models, and unless there is a compelling reason to believe otherwise, it is appropriate to assume the generic case. For example, energy conserving models of the form $\ddot{p} + f(p) = 0$ lead to linearizations in which the equilibria are non-hyperbolic. If one is willing to accept that energy dissipation in such models can be ignored, then we are irretrievably led to accept eigenvalues having zero real parts. However if, as some argue, dissipation is physically inevitable, then it is possible to concede the hyperbolicity of equilibria without serious pangs of conscience.

3.3 Feedback

Let us take another look at the pendulum (Section 2.1). Recall that its upright position (at $\theta = \pi, \dot{\theta} = 0$) is an unstable equilibrium. However, what if somehow we could move the pivot laterally? This would provide

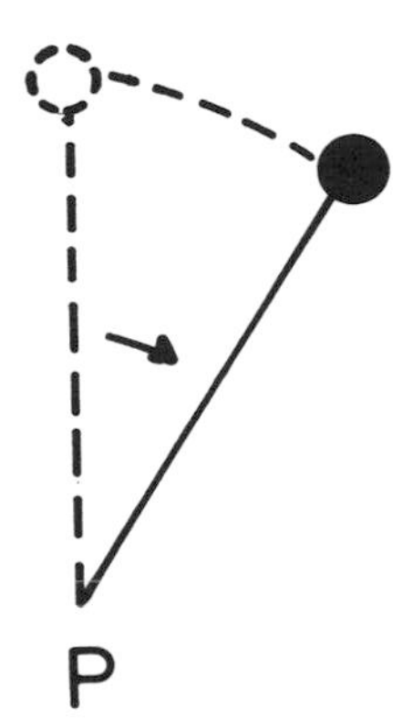

Figure 3.4 Movement away from the unstable position of the pendulum.

an additional degree of motion, which could be used to keep the bob upright. To be specific, consider the bob disturbed slightly from its unstable equilibrium (Figure 3.4). Now move the pivot center at P rapidly to the right by an amount w (Figure 3.5). The angle of declination away from the straight up position is less after moving P than before.

Although θ previously measured the counterclockwise angle of movement from the straight down position of the bob, we now let it denote the clockwise direction away from straight up. Suppose that some sensor (our eye, for instance) measures the angle θ and velocity $\dot{\theta}$ at each instant of time and then immediately feeds this information to some

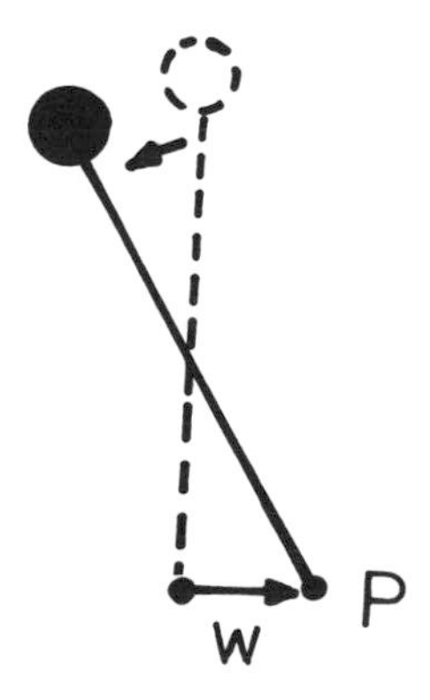

Figure 3.5 Movement away from the unstable position of the pendulum after the pivot at P is quickly shifted to the right by an amount w.

correcting device (our hand). This device, the controller, uses the information to move the pivot either right or left. The hand movement induces a restoring force v, which depends on θ and $\dot{\theta}$, and it is called a *feedback control*. As the sensor data changes, so does v.

Generally speaking, a control is externally imposed on a given system in an attempt to modify any undesirable behavior. In our case, we hope to keep the bob upright. Can it be done? Certainly if the unstable position is stabilized by some appropriate feedback control v, the answer is yes, but the problem is to show that the modified pendulum does indeed possess a stable equilibrium. To understand more clearly what is happening, it is first necessary to rewrite the equations of motion so as to include the control. Incidently, since it is sort of silly to talk about a pendulum with a movable pivot, let us interpret the problem as that of balancing a thin stick upright on the palm of a hand, with the hand moving laterally. At one time or other most of us have tried to do this, and so the question of how to move our hand so as to maintain balance is certainly plausible. To be consistent with the pendulum problem assume that all the mass of the stick is located at a point on its upper end.

The angle θ is measured clockwise from the vertical. As in Section 2.1, the bob moves along a circular arc where distance is designated by $s = l\theta$. The quantity l is the stick length. There are three forces on the mass m at the end of the stick. One is a downward gravitational pull, another is a viscous damping resistance to motion (both of these familiar from the pendulum), and there is also a lateral restoring force due to the control (Figure 3.6). It is necessary to resolve these forces into components which are, at each instant, along the actual direction of motion, namely along the tangent to the circular arc of radius l. Observe that the

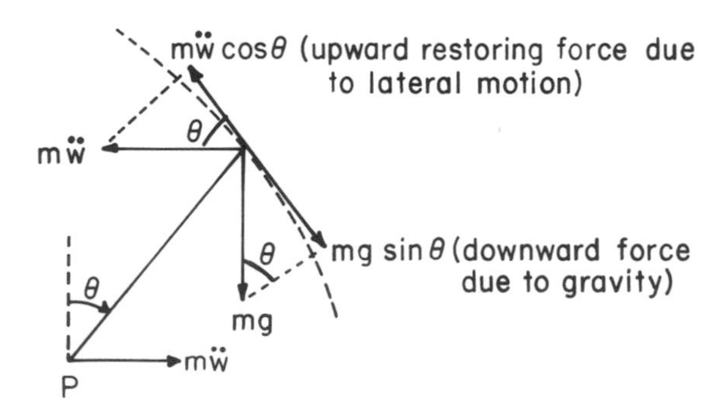

Figure 3.6 Force diagram for the inverted pendulum.

hand's lateral motion exerts a force on the mass in the opposite direction. An acceleration of the hand to the right moves the pivot left. By Newton's law, the restoring force is therefore $v = -m\ddot{w}$.

The total force on the mass is then

$$m\ddot{s} = mg\sin\theta - k_1\dot{s} - m\ddot{w}\cos\theta.$$

As before, the damping force is proportional to velocity $\dot{s}$ along the arc. Since $\ddot{s} = l\ddot{\theta}$ and $\dot{s} = l\dot{\theta}$, the differential equation can be written (with $r = k_1/m$) as

$$\ddot{\theta} = \frac{g}{l}\sin\theta - \frac{\ddot{w}}{l}\cos\theta - r\dot{\theta}. \tag{3.22}$$

The negative signs indicate forces in a direction opposite the positive θ sense.

As usual, rewrite 3.22 as a first-order system by letting $x_1 = \theta$, $x_2 = \dot{\theta}$. Then

$$\begin{aligned}\dot{x}_1 &= x_2\\ \dot{x}_2 &= -rx_2 - \frac{\ddot{w}}{l}\cos x_1 + \frac{g}{l}\sin x_1.\end{aligned} \tag{3.23}$$

At this point we make a modeling assumption, which is that the control v depends linearly on the position angle θ and its velocity $\dot{\theta}$. We do this for mathematical convenience in the hope that it is not too restrictive a condition. Thus,

$$v = c_1x_1 + c_2x_2 \tag{3.24}$$

for suitable constants c_1, c_2, which remain to be determined. Substituting 3.24 into 3.23 gives

$$\begin{aligned}\dot{x}_1 &= x_2\\ \dot{x}_2 &= -rx_2 + \frac{c_1x_1}{l}\cos x_1 + \frac{c_2x_2}{l}\cos x_1 + \frac{g}{l}\sin x_1.\end{aligned} \tag{3.25}$$

An equilibrium occurs at $x_1 = x_2 = 0$, which is the straight up position at which $\theta = \dot{\theta} = 0$.

Our goal is to choose v at each instant so that the equilibrium $\bar{\mathbf{x}} = \mathbf{0}$ becomes asymptotically stable. This would mean that to an observer who is unaware that any control is being exerted, the upright position appears to be miraculously stable: small movements away from the vertical tend to be compensated for and the stick teeters about its equilibrium until it settles down. This improbable motion is the result of feedback. The successful use of feedback control underlies much of contemporary engineering.

The system 3.25 can be linearized, and if the resulting linear system is asymptotically stable then, at least locally, so is 3.25.

The linearized equations about $\bar{\mathbf{x}} = \mathbf{0}$ (Exercise 3.4.6) result in

$$\begin{aligned} \dot{u}_1 &= u_2 \\ \dot{u}_2 &= \left(\frac{g}{l} + \frac{c_1}{l}\right)u_1 + \left(\frac{c_2}{l} - r\right)u_2. \end{aligned} \tag{3.26}$$

The variables u_i are actually identical to x_i in this case since $\bar{\mathbf{x}} = \mathbf{0}$. It is apparent that if $c_1 = -(l + g)$ and $c_2 = rl - l$, then 3.26 becomes

$$\begin{aligned} \dot{u}_1 &= u_2 \\ \dot{u}_2 &= -u_1 - u_2. \end{aligned} \tag{3.27}$$

This is a system of the form $\dot{\mathbf{u}} = A\mathbf{u}$ with

$$A = \begin{pmatrix} 0 & 1 \\ -1 & -1 \end{pmatrix}.$$

Trace $A < 0$ and Det $A > 0$, and therefore the equilibrium is asymptotically stable, as required (Lemma 2.2). Thus, a hand movement that is proportional to the observed θ and $\dot{\theta}$ values suffices to keep the stick balanced.

Since $\mathbf{u}$ is the same as $\mathbf{x}$ anyway (at least in our example), if we rewrite 3.26 in terms of $\mathbf{x}$ and use the fact that $v/m = c_1 x_1 + c_2 w_2$, then we get

$$\begin{aligned} \dot{x}_1 &= x_2 \\ \dot{x}_2 &= \frac{g x_1}{l} - r x_2 + \frac{v}{ml}. \end{aligned} \tag{3.28}$$

In vector notation 3.28 becomes

$$\dot{\mathbf{x}} = A\mathbf{x} + v\mathbf{b}, \tag{3.29}$$

where

$$\mathbf{b} = \begin{pmatrix} 0 \\ \dfrac{1}{ml} \end{pmatrix} \quad \text{and} \quad A = \begin{pmatrix} 0 & 1 \\ \dfrac{g}{l} & -r \end{pmatrix}.$$

The particular form of A in 3.29 is the result of starting with a scalar second-order differential equation model. Often, however, nonlinear models lead to linearized systems of the form 3.29 in which A has the more general form

$$A = \begin{pmatrix} a_{11} & a_{12} \\ a_{21} & a_{22} \end{pmatrix}, \qquad \mathbf{b} = \begin{pmatrix} 0 \\ 1 \end{pmatrix}.$$

If one assumes a linear feedback law 3.24 or, equivalently,

$$v = \mathbf{c} \cdot \mathbf{x}, \tag{3.30}$$

where the dot is scalar or dot product of vectors and $\mathbf{c} = \begin{pmatrix} c_1 \\ c_2 \end{pmatrix}$, then 3.29 reduces to

$$\dot{\mathbf{x}} = (A + \mathbf{b}\mathbf{c}^{\mathrm{T}})\mathbf{x}. \tag{3.31}$$

The zero equilibrium of 3.31 is asymptotically stable if $\mathbf{c}$ can be chosen so that the eigenvalues of the matrix $A + \mathbf{b}\mathbf{c}^{\mathrm{T}}$ have negative real parts. The following lemma tells us when this is possible and is known as the *eigenvalue placement theorem* in the control theory literature.

Lemma 3.1 *Suppose that the matrix B with columns $\mathbf{b} = \begin{pmatrix} 0 \\ 1 \end{pmatrix}$ and $A\mathbf{b}$ is nonsingular. Then a vector $\mathbf{c}$ can be chosen so that the zero solution of* 3.31 *is asymptotically stable.*

Proof An easy computation shows that

$$B = \begin{pmatrix} 0 & a_{12} \\ 1 & a_{22} \end{pmatrix}.$$

Therefore, B is invertible if $a_{12} \neq 0$. Also

$$A + \mathbf{b}\mathbf{c}^{\mathrm{T}} = \begin{pmatrix} a_{11} & a_{12} \\ a_{21} + c_1 & a_{22} + c_2 \end{pmatrix},$$

so that

$$\begin{aligned} \mathrm{Trace}(A + \mathbf{b}\mathbf{c}^{\mathrm{T}}) &= \mathrm{Trace}\, A + c_2 \\ \mathrm{Det}(A + \mathbf{b}\mathbf{c}^{\mathrm{T}}) &= \mathrm{Det}\, A + c_2 a_{11} - c_1 a_{12}. \end{aligned}$$

It is not difficult to see that since $a_{12} \neq 0$, then c_1 and c_2 can be chosen to make the trace negative and the determinant positive. By Lemma 2.1, this proves the result. Observe that if $a_{12} = 0$, there is no guarantee this could be done.

The technical condition that the matrix B in Lemma 3.1 is nonsingular is a special case of a property known as *controllability*. It plays an important role in the study of certain dynamic systems in engineering.

A connection between feedback controls and Liapunov functions will be uncovered in Chapter Nine when we establish that such controls can also be optimal in a sense to be made precise later.

3.4 Exercises

3.4.1 Let V be a Liapunov function on a set Ω in which $\bar{\mathbf{x}}$ is an isolated equilibrium. Show that $V(\bar{\mathbf{x}}) = 0$.

3.4.2 Show that the function 3.7 is a Liapunov function in Example 3.3 of the text, and obtain an estimate of Ω_0. Explain why 3.7 is a reasonable "energy function".

3.4.3 Show that the origin in the damped mass–spring system (Equation 3.10) is globally a.s. Use Theorem 3.1.

3.4.4 In doing isocline analysis (see Section 3.3), it is sometimes useful to observe that if $\dot{x}_1 \neq 0$, then

$$\frac{dx_2}{dx_1} = \frac{\dot{x}_2}{\dot{x}_1} = \frac{f_2(x_1, x_2)}{f_1(x_1, x_2)}.$$

Show why this is true using the chain rule of calculus. This differential equation can be used to infer how x_2 varies as a function of x_1 along an orbit in the x_1, x_2 plane. In fact, dx_2/dx_1 gives the direction of an orbit since it is the slope of the solution curve at all points where $\dot{x}_1$ is nonzero (that is, where the orbit is not vertical). Use this idea to redo the isocline analysis of the damped mass–spring in Example 3.5.

3.4.5 The only equilibrium of the system 3.17 is the origin. Show this.

3.4.6 Establish that the linearized system corresponding to 3.25 in the text is 3.26.

3.4.7 Reformulate Exercise 1.5.7 by letting the rate at which water enters the tank be proportional to the depth of water h. Show that the nontrivial equilibrium $\bar{h}$ is unstable and explain why this should be so physically.

3.4.8 Stabilize the unstable equilibrium $\bar{h}$ of the previous exercise by adding a control that depends linearly on the deviation of h from $\bar{h}$. Solve this feedback control problem using the ideas of Section 3.3 in the text.

3.4.9 It is generic for a square matrix to be nonsingular. Establish this for 2×2 matrices.

Chapter Four

Growth and Decay

4.1 The Logistic Model

A population may consist of people, animals, insects, bacteria, or other self-reproducing organisms. We wish to model its growth over time. Reproduction is assumed to be carried out continuously by all its members, without regard to age or sex differences, at an average per-capita rate r. This constant is the net difference between per-capita birth and death rates averaged over the entire population. The population level at time t is denoted by $N(t)$.

Although N is integer valued, it is convenient to think of it as a smooth function. For large N, the distinction between a continuous or integer-valued population becomes blurred and so the smoothness requirement is not a serious restriction. In later chapters, $N(t)$ is sometimes treated as a population density (per unit area), which is generally noninteger. However, some of our other assumptions may be questionable and a few of them will be shed as we go along.

To begin with, consider the environment inhabited by our population to be closed in the sense of no migration into or out of the space it occupies. Since each member of the population reproduces at the same per-capita rate r, the total growth rate is rN:

$$\dot{N} = rN. \tag{4.1}$$

This simple equation is separable and has the solution

$$N(t) = N(0)e^{rt}, \tag{4.2}$$

and so the population apparently grows without bound as t increases. This is unrealistic for large t, but within more modest time frames, 4.2 often gives acceptable results. Examples are given in the references cited.

The reason for the eventual breakdown of 4.2 as t increases is not hard to see. With N large, one expects the competition for living space and scarce resources to lead to conflict while overcrowding can result in epidemics and disease. Each of these factors tend to limit the per-capita growth rate.

In order to improve the model, suppose the habitat can support a maximum population level K. This is known as the *carrying capacity*. When N reaches K, the growth rate is taken to be zero. This represents the extreme case in which the capacity for growth has been saturated. A reasonable modification of r to account for this limited capacity is to consider a per-capita growth rate that decreases as N increases. A plausible candidate is

$$r\left[1 - \frac{N(t)}{K}\right]. \tag{4.3}$$

Using 4.3 in place of r in 4.1 gives what we hope will be a better growth model:

$$\dot{N} = rN\left(1 - \frac{N}{K}\right). \tag{4.4}$$

This is called the *logistic equation*. Although nonlinear, it can be integrated by separating variables. However, it is instructive to see first how much we can learn about its solutions by other means in the spirit of the previous chapters. Since $\dot{N} \geq 0$ for N between its equilibrium values of zero and K, therefore over time it moves asymptotically toward K unless it is at one of these rest points to begin with. Thus, zero is an unstable state while K is a.s. Observe that $N(t)$ identically equal to K is a solution of 4.4. Therefore, no orbit can exceed the value K if it starts from below or on this value because of the assumed uniqueness of orbits.

An alternate approach to stability is obtained by using Taylor's theorem from calculus (it may be worthwhile to refer to the previous use of this theorem in the case of two variables that was given in Section 2.3). Let $\bar{N}$ be an equilibrium point of 4.4 (either zero or K), and define a small perturbation away from $\bar{N}$ by

$$u(t) = N(t) - \bar{N}.$$

Then the first-order Taylor's theorem gives, approximately,

$$f(N) = f(\bar{N}) + \frac{df(\bar{N})}{dN} u.$$

If we take $f(N) = rN(1 - N/K)$, then $f(\bar{N}) = 0$. Also $\dot{u} = \dot{N}$, and so the linearized equation corresponding to 4.4 is

$$\dot{u} = r\left(1 - \frac{2\bar{N}}{K}\right)u. \tag{4.5}$$

When $\bar{N} = 0$, 4.5 leads to

$$\dot{u} = ru,$$

whose solution increases exponentially for $r > 0$. This shows that zero is indeed unstable. On the other hand, $\bar{N} = K$ gives

$$\dot{u} = -ru,$$

and in this case u tends to zero as t increases so that K appears to be a.s. These qualitative insights are completely vindicated by solving 4.4 explicitly. Separating variables in the usual way one obtains

$$K \int \frac{dN}{N(K-N)} = rt + \text{constant}.$$

The left side of this relation may be written as

$$\int \frac{dN}{N} + \int \frac{dN}{(K-N)} = \ln\left(\frac{N}{K-N}\right),$$

and, for a suitable constant c, we arrive at the solution

$$N(t) = \frac{K}{1 + ce^{-rt}}. \tag{4.6}$$

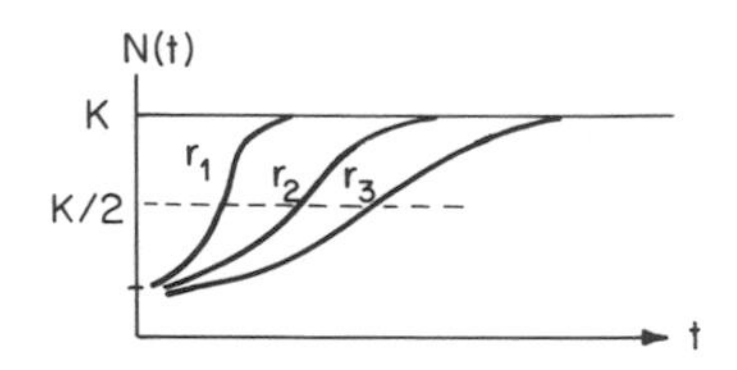

Figure 4.1 Solutions of Equation 4.4 for $r_1 < r_2 < r_3$.

$N(t)$ goes to K as expected provided that the orbit is not initially at a rest point. The plot of $N(t)$ has a characteristic "S" shape indicating rapid growth for $N < K/2$ (it is easy to check that $K/2$ is an inflection point of 4.6) with slower growth thereafter (Figure 4.1). This indicates the effect of overcrowding and competition as the population increases, and it gives results that are qualitatively closer to what one actually observes than does the simpler model 4.1. An oft-quoted example is that of microscopic paramecium growing in a test tube in which there is a limited amount of nutrient. Their growth curve is very much logistic.

If migration is permitted, let $E > 0$ denote the constant per-capita outflow. For example, if N is the population of some country and five out of every hundred people emmigrate each day then E would be .05. Or if N denotes the level of fish in a lake and the catch rate is one out of ten fish then E is .10. Because outflow diminishes the population, 4.4 must be modified:

$$\dot{N} = rN\left(1 - \frac{N}{K}\right) - EN. \tag{4.7}$$

The negative sign indicates outflow. With E negative, one has an equivalent model of inflow (immigration or restocking of the lake). It is left as an exercise to solve and interpret this model (Exercise 4.5.1).

We have assumed that per-capita growth decreases with increasing N due to overcrowding and competition. It is conceivable, however, that a population is affected by its environment in quite different ways. Fish that swim in schools, for instance, seem to be better protected from predators than when they swim alone. This safety-in-numbers idea suggests that per-capita growth could increase with N for small values of the population since a bigger crowd enhances the chances of survival. As N gets large enough, however, the effects of overcrowding begin to be felt and the growth rate starts to decrease. Let $f(N)$ be the growth rate. An

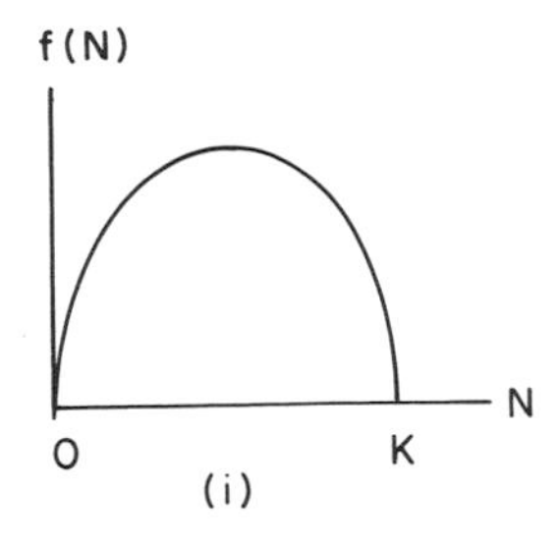

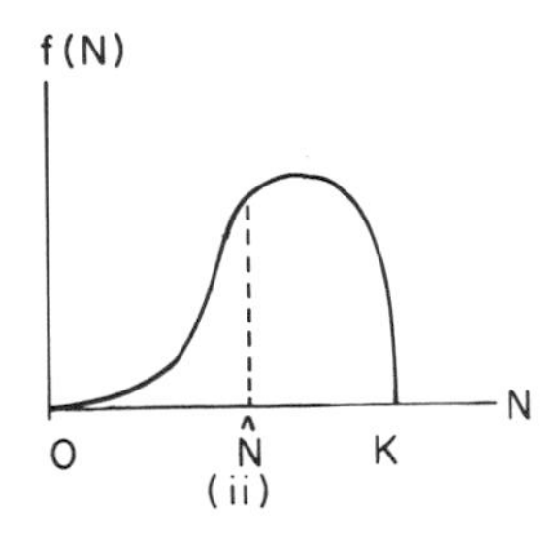

Figure 4.2 Growth rates in which $f(N)/N$ (i) decreases with N (logistic growth) and (ii) increases with N for $N < \hat{N}$.

explicit functional form is postponed to Chapter Six. All that we need assume for now is that the per-capita rate $f(N)/N$ increases with N for N less than some value $\hat{N}$ and decreases thereafter. Equation 4.4 then generalizes to

$$\dot{N} = f(N). \tag{4.8}$$

A typical situation is displayed in Figure 4.2 where it is compared to the logistic case.

Another interesting possibility is where scarcity in numbers is a detriment. When the habitat is large, small numbers could imply eventual extinction since there are not enough individuals to mate with. Could this be the fate of certain whale species?

In order to model this gloomy scenario, assume a growth rate that is negative for N less than some critical value $\hat{N}$. A typical case is sketched in Figure 4.3 in which growth is logistic only for $N > \hat{N}$. From this picture what would you conclude about stability?

Let us now remove another restriction from the logistic model. Allow K to vary over time. Suppose, for example, that a population's growth

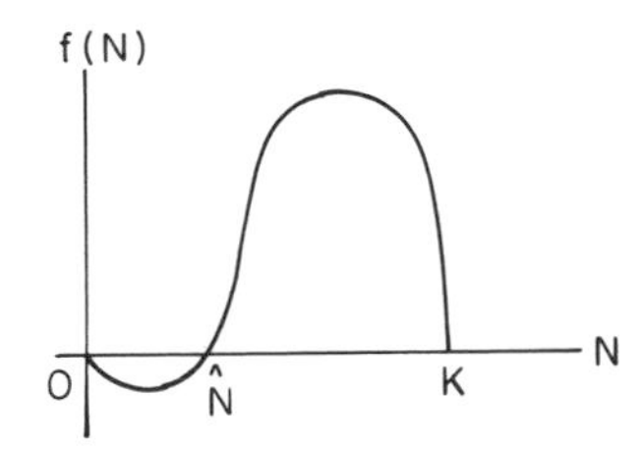

Figure 4.3 Growth rate in which $f(N)/N$ is negative for $N < \hat{N}$.

capacity fluctuates with the seasons because of differences in sunlight and temperature. Then K would be a roughly periodic function with a period of one year. With K variable, 4.4 becomes

$$\dot{N}(t) = rN(t)\left[1 - \frac{N(t)}{K(t)}\right]. \tag{4.9}$$

This is a special case of a nonlinear equation of the form

$$\dot{y} = a(t)y - b(t)y^2, \tag{4.10}$$

which is known as *Bernoulli's equation*. A nice trick that permits one to solve 4.10, and hence 4.9, is to introduce a new variable $x = 1/y$. Substituting into 4.10 gives

$$x\dot{y} + \dot{x}y = 0. \tag{4.11}$$

In this manner, we obtain a linear first-order and non-homogeneous equation in x:

$$\dot{x}(t) + a(t)x(t) - b(t) = 0. \tag{4.12}$$

In the special case of 4.9, this reduces to the equation

$$\dot{x}(t) + rx(t) - \frac{r}{K(t)} = 0. \tag{4.13}$$

Of course we are interested in N, which is simply the reciprocal of x.

As it stands, 4.13 is a bit too general to be really useful. A good modeling ploy is to introduce plausible assumptions which simplify its solution until it becomes tractable. However, we will not consider it further.

4.2 Discrete Versus Continuous

Reproduction has, up to now, been continuous in time. However, certain organisms reproduce at discrete intervals. Insects such as the gypsy moth reproduce once a year, for example. In such cases model 4.4 is misleading. A more suitable formulation leads to a nonlinear difference equation. If $N(n\,\Delta t)$ is the population at the nth time interval of duration Δt, then the value of N at the $(n + 1)$th interval is found by analogy with the way 4.4 was obtained. Therefore,

$$N((n+1)\,\Delta t) = N(n\,\Delta t) + rN(\Delta t)\left[1 - \frac{N(n\,\Delta t)}{K}\right]\Delta t. \quad (4.14)$$

Observe that by bringing $N(n\,\Delta t)$ to the left side of 4.14, dividing by Δt, and then letting Δt tend to zero, one obtains 4.4 in the limit (assuming that N is differentiable). However, as models go, 4.14 is very different in its behavior than its smooth cousin, and one must be careful to use either 4.4 or 4.14 to indicate continuous or discrete growth depending on which is meant. In order to see why this is so, consider the solution $N(n\,\Delta t) = K$ of 4.14. It represents an equilibrium since it implies no further change in growth. Let us examine its stability. To do this, we mimic the linearization procedure used for differential equations. Simply let

$$u(n\,\Delta t) = N(n\,\Delta t) - K, \quad (4.15)$$

and assume an initial perturbation $u(0) = N(0) - K$ that is "small". If $u(n\,\Delta t) \to 0$ as $n \to \infty$, then we say that K is a.s. Substituting 4.15 into 4.14 gives, after cancellations,

$$u((n+1)\,\Delta t) = u(n\,\Delta t) - \frac{r\Delta t}{K}[u(n\,\Delta t) + K]u(n\,\Delta t). \quad (4.16)$$

But, since $u(n\,\Delta t)$ is small, $[u(n\,\Delta t)]^2$ is negligible. This leads to the approximate linear difference equation

$$u((n+1)\,\Delta t) = u(n\,\Delta t)(1 - r\Delta t). \quad (4.17)$$

Setting $n = 0, 1, 2, \ldots,$ one obtains recursively from 4.17 that

$$\begin{aligned} u(\Delta t) &= u(0)(1 - r\Delta t) \\ u(2\,\Delta t) &= u(0)(1 - r\Delta t)^2 \\ &\vdots \\ u(n\,\Delta t) &= u(0)(1 - r\Delta t)^n. \end{aligned} \quad (4.18)$$

It is now apparent that $u(n\,\Delta t)$ tends to zero if and only if

$$|1 - r\Delta t| < 1. \quad (4.19)$$

Or, equivalently, if and only if

$$0 < r\,\Delta t < 2. \tag{4.20}$$

When $r\,\Delta t = 2$, then $u(n\,\Delta t) = (-1)^n u(0)$, and so u oscillates back and forth. This represents a stable cycle and is called neutral stability. It corresponds to the case of purely imaginary eigenvalues for linear differential equations. Since $N(n\,\Delta t) = K + u(0)$, it signifies that the population jumps back and forth in alternate time periods between two values.

If $r\,\Delta t > 2$, the population is not restored to equilibrium. This is in contrast with the continuous model 4.4 in which the rest state K is a.s. regardless of the value of r. So the discrete model is quite sensitive to the per-capita growth rate as well as the interval length between reproduction. Actually, this is only the beginning since N can exhibit unexpectedly bizarre behavior as r increases. This intriguing story will be taken up again later in Section 8.3.

4.3 The Struggle for Life, I

We have taken a look at the dynamics of growth of a single population. These ideas are now extended to include two interacting populations whose levels at time t are $N_1(t)$ and $N_2(t)$. The models to be discussed follow the work of Volterra and Lanchester, among others, and are now classics. There are numerous expositions of this work, and we confine ourselves to a rapid overview of a few among the many variants that are possible.

Implicit in each model is that there is a conservation law at work. If one ignores the fact that populations can reproduce themselves (which is a source term) and can also die (a sink term), the net change in the population within a closed environment can only be due to the flow across environmental boundaries. This is a paraphrase of the conservation of mass principle enunciated in Section 1.4. By adding in the sources and subtracting out the sinks one is then able to write down equations for the rate of change of population levels.

Example 4.1 (Predator–Prey) Suppose a species P_1 feeds on an unlimited food source. It is called the prey. Another species P_2 feeds on P_1 and is called the predator. When P_1 is plentiful, P_2 is able to increase in numbers, but in the absence of P_1 the predator dies out (it is assumed that P_1 is the sole food source of P_2). Let us suppose that if P_1 is left to itself without predation, it grows logistically, whereas P_2, without access to P_1, decreases at a constant per-capita rate $c > 0$. In the presence of mutual interaction, however, the growth rate of the prey must diminish while that of the predator increases. To model the interaction term, one supposes that encounters between the two species occurs as a sort of "hide and seek" game and is proportional to the total number of possible ways in which P_1 and P_2 can meet. That is, interaction is proportional to the product of the population levels $N_1 N_2$. Putting these plausible assumptions together into a single package results in the model

$$\begin{aligned} \dot{N}_1 &= rN_1\left(1 - \frac{N_1}{K}\right) - \alpha N_1 N_2 \\ \dot{N}_2 &= -cN_2 + \beta N_1 N_2. \end{aligned} \tag{4.21}$$

The positive constants of proportionality α and β represent efficiencies: α is the per-capita rate at which the predator captures its prey, while β is the per-capita growth rate of the predator resulting from its catch.

The model 4.21 assumes a closed habitat involving only two species with no outside interference. In practice other factors intervene. Some prey can hide from their predators, predators may have other sources of food available to them, and predators may not be able to feed on unlimited amounts of food. In this last case, one must model the effects of satiation. These and other modifications are discussed in the exercises and in Chapter Six.

Example 4.2 (Competition) A related class of models concerns competition in which two populations P_1 and P_2 vie for the same food source. Both species are assumed to grow logistically in isolation from each other, but when they interact, each interferes with the growth of the other. The same kind of interaction term as in the predator–prey case has the effect of reducing the growth rates of P_1 and P_2 as they both struggle for the same commodity. This leads to the pair of equations

$$\begin{aligned} \dot{N}_1 &= rN_1\left(1 - \frac{N_1}{K}\right) - \alpha N_1 N_2 \\ \dot{N}_2 &= sN_2\left(1 - \frac{N_2}{L}\right) - \beta N_1 N_2, \end{aligned} \tag{4.22}$$

in which L is the carrying capacity of P_2. The positive constants α and β now represent per-capita rates of competitive advantage of one species over the other. Interpret $\alpha > \beta$ to mean that P_2 is the more efficient competitor, and vice-versa for $\beta > \alpha$. There are numerous examples of competition and predation in nature: one has only to think of two strains of yeast feeding on the same nutrient or of small crustaceans grazing on algae.

Example 4.3 (Combat) Combat represents a specific type of competition between humans. Suppose now that P_1 and P_2 are opposing forces. Under conditions of guerilla warfare, which resembles the hide and seek situation discussed earlier, each side suffers losses in proportion to the number of different encounters that are possible, namely $N_1 N_2$. Moreover each side, being human, also suffers attrition due to accidents, disease, and even dissertion. These losses are assumed proportional to the level of troop size. Putting this together gives the model

$$\begin{aligned} \dot{N}_1 &= -aN_1 - \alpha N_1 N_2 \\ \dot{N}_2 &= -bN_2 - \beta N_1 2 N_2. \end{aligned} \tag{4.23}$$

All signs are negative (the constants are positive) since P_1 and P_2 inevitably diminish over time without outside reinforcements, which are here excluded.

In conventional combat, in which both forces are in view of each other, the losses due to the enemy are proportional to the size of the enemy force. This is because the chance of hitting a member of side 1, let us say, is dependent only on how many troops there are on side 2. Assuming all troops have the same kill capability, then the larger the troop size, the more deadly it is. After eliminating a member of side 1, firepower is drawn to another member of side 1, and so the size of force 1 is irrelevant. So, for conventional combat, the model becomes

$$\begin{aligned} \dot{N}_1 &= -aN_1 - \alpha N_2 \\ \dot{N}_2 &= -bN_2 - \beta N_1. \end{aligned} \tag{4.24}$$

It would be possible to give numerous other models that lead to similar equations. Some of these are found in the exercises. Instead, our attention now turns to understanding something of what such models tell us.

All the models considered in this section are special cases of a general expression involving at most quadratic terms in N_1 and N_2. For this reason, it is called the *quadratic model*:

$$\begin{aligned}\dot{N}_1 &= N_1(a_1 N_1 + b_1 N_2 + d_1)\\ \dot{N}_2 &= N_2(a_2 N_1 + b_2 N_2 + d_2)\end{aligned} \tag{4.25}$$

It is difficult to say very much about the quadratic model, but at least one important property emerges right away. Observe, first, that the coordinate axis represent solutions of 4.25. For example, if $N_1 = 0$, then a solution exists on the vertical axis for any initial value of N_2. The same is true on the horizontal axis when the roles of N_1 and N_2 are reversed. It follows that if $\mathbf{N} = \begin{pmatrix} N_1 \\ N_2 \end{pmatrix}$ starts in the positive quadrant, then because of the uniqueness of orbits, no solution of 4.25 can cross the coordinate axis, and so it remains in this quadrant. This is important from a modeling viewpoint because N_1 and N_2 represent populations and, as such, should remain nonnegative for all time.

A stability analysis of 4.25 is carried out in the next section. For now let us see what a simple isocline approach reveals in the N_1, N_2 plane. One case will suffice, and we choose the competition model 4.22. There are evidently four possible equilibria:

$$\mathbf{N} = \begin{pmatrix} 0 \\ 0 \end{pmatrix}, \begin{pmatrix} K \\ 0 \end{pmatrix}, \begin{pmatrix} 0 \\ L \end{pmatrix}, \begin{pmatrix} \overline{N}_1 \\ \overline{N}_2 \end{pmatrix}.$$

In the last case, $\overline{\mathbf{N}}$ satisfies

$$\begin{aligned}\overline{N}_1 &= \frac{s}{\beta} - \frac{s}{\beta L}\overline{N}_2\\ \overline{N}_2 &= \frac{r}{\alpha} - \frac{r}{\alpha K}\overline{N}_1.\end{aligned} \tag{4.26}$$

The isoclines for the model 4.22 are all straight lines (Figure 4.4).

Consider the situation pictures in Figure 4.4b. Fix a value of N_1, and suppose N_2 is above the line defined by $\dot{N}_1 = 0$. A glance at Equations 4.22 shows that $\dot{N}_1 < 0$. The reverse holds if N_2 is below the line. Similarly, if N_2 is above the isocline $\dot{N}_2 = 0$, then $\dot{N}_2 < 0$, with the opposite being true when N_2 is below. Combining this information provides us with Figure 4.5.

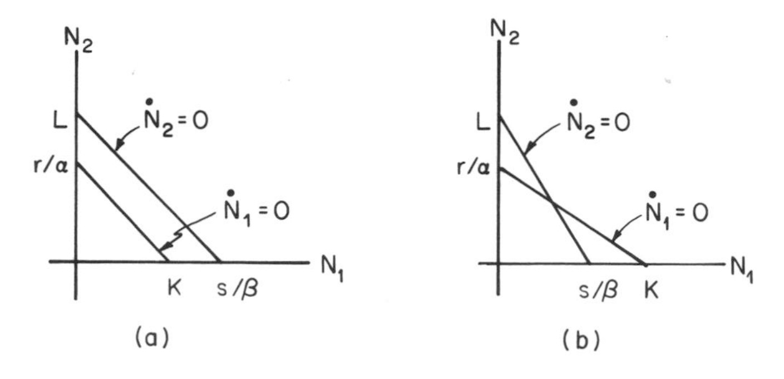

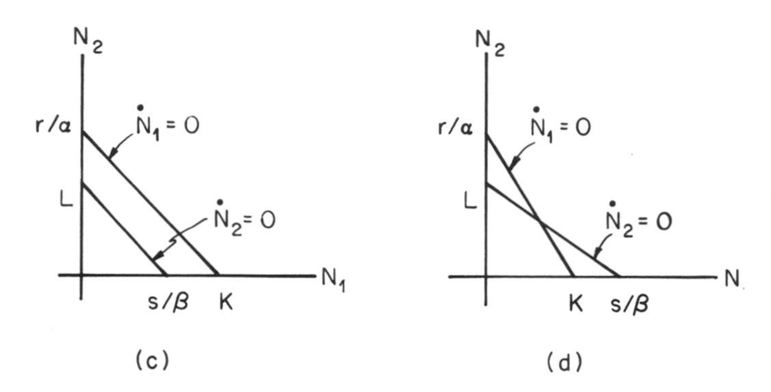

Figure 4.4 Isoclines for the competition model 4.22.

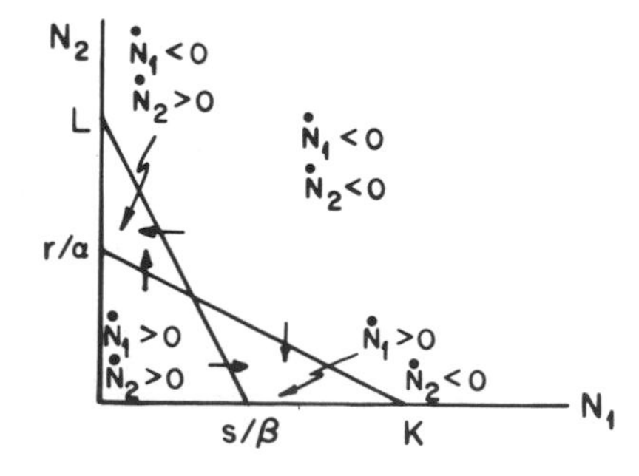

Figure 4.5 Isocline analysis for case (b) in Figure 4.4.

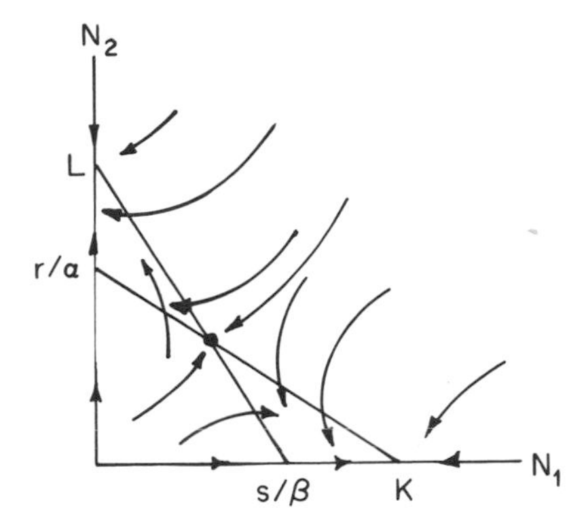

Figure 4.6 Orbits for model 4.22 in the case (b) of Figure 4.4 (strong competition).

When an orbit of 4.22 crosses an isocline, it does so either vertically or horizontally since either $\dot{N}_1$ or $\dot{N}_2$ is zero there. Moreover with $N_1 = 0$, one sees that $\dot{N}_2 > 0$ if $N_2 < L$ and $\dot{N}_2 < 0$ for $N_2 > L$. A similar result holds when $N_2 = 0$. We now have sufficient information to sketch the orbit directions as in Figure 4.6. In doing so, keep in mind that the orbits cross the isoclines smoothly in the indicated vertical and horizontal directions. This tells us how the arrows in Figure 4.6 must bend.

In the present example, it is fairly apparent that the nontrivial equilibrium is a saddle and hence unstable. This suggests that except for the stable manifold of $\overline{\mathbf{N}}$, all orbits are separated into those that tend either to $\begin{pmatrix} 0 \\ L \end{pmatrix}$ or $\begin{pmatrix} K \\ 0 \end{pmatrix}$. Depending on the initial starting values of N_1 and N_2, one or the other population eventually dies out. This is sometimes known as the *principle of competitive exclusion* in ecological modeling since it suggests that two strongly competing species cannot coexist within the same habitat.

To understand what strongly competitive means, consider the special case in which the growth rates and carrying capacity are identical ($r = s$, $K = L$). When the populations are nearly equal in size ($N_1 = N_2$), the self-limitation factor in the logistic term is $-N_1^2 r/K$, whereas the factor that expresses the limitation due to competition is $-\alpha N_1^2$, or $-\beta N_2^2$. However, from Figure 4.4b, we see that α and β each exceed r/K suggesting, for example, that one of them consumes food more voraciously than the other, or it leaves wastes that are toxic to the other, or is otherwise successful in intimidating its opponent. Such fierce interaction eliminates one of the species or the other. However, in the case of Figure 4.4d just the opposite holds. Here competition is mild since α and β are smaller and so the equilibrium is now a.s. Coexistence is possible. The other two cases can be analyzed in a similar fashion.

4.4 Stable Equilibria, III

All the models formulated in the previous section are also susceptible to a linearized stability analysis. Since the arguments are by now familiar, let us consider only one of the models briefly, the predator–prey 4.21. Other than the uninteresting equilibria in which one or the other species is zero, the remaining nontrivial rest point occurs at

$$\begin{aligned} \overline{N}_1 &= \frac{c}{\beta} \\ \overline{N}_2 &= \left(\frac{r}{\alpha}\right)\left(1 - \frac{c}{\beta K}\right). \end{aligned} \tag{4.27}$$

Since $\overline{N}_1$ cannot exceed the maximum population level K, it follows that $\overline{N}_2 > 0$. The linearized system about this point satisfies

$$\dot{\mathbf{u}} = A\mathbf{u},$$

in which

$$A = \begin{pmatrix} -\frac{r}{K}\overline{N}_1 & -\alpha\overline{N}_1 \\ \beta\overline{N}_2 & 0 \end{pmatrix}.$$

It is apparent that Trace $A < 0$ and Det $A > 0$. By Lemma 2.2, the equilibrium 4.27 is a.s.

Now modify the model by introducing the effect of some external agent that indiscriminantly harvests both species at the same per-capita rate $E > 0$. Two examples will be given presently, but first let us write down equations in which the original growth rates of 4.21 are reduced by the harvest rates $-EN_1$ and $-EN_2$:

$$\begin{aligned} \dot{N}_1 &= rN_1\left(1 - \frac{N_1}{K}\right) - \alpha N_1 N_2 - EN_1 \\ \dot{N}_2 &= -cN_2 + \beta N_1 N_2 - EN_2. \end{aligned} \tag{4.28}$$

The nontrivial equilibrium of 4.28 is

$$\hat{N}_1 = \overline{N}_1 + \frac{E}{\beta}$$
$$\hat{N}_2 = \overline{N}_2 - \frac{E}{\alpha}\left(\frac{r}{\beta K} + 1\right). \tag{4.29}$$

From 4.29, it is seen that $\hat{N}_1 > \overline{N}_1$ and $\hat{N}_2 < \overline{N}_2$, which means that harvesting changes the stable equilibrium in such a way that there are now more prey and less predators.

An oft-quoted example is that of sharks in the Adriatic Sea. They act as predators on other kinds of fish. During World War I, when fishing in the Adriatic was temporarily reduced, it was observed that the percentage of sharks caught by the fishermen actually increased. A plausible explanation is provided by the model: 4.29 shows that as the fishing rate E decreases, the quantity $\hat{N}_1$ would decline while $\hat{N}_2$ gets bigger.

Another illustration of this idea comes from considering a pest which attacks the orange groves in California. The pest has a natural predator, a kind of lady bug, which controls the total pest population. A pesticide spray is applied to the grove. It kills both predator and prey, and so we would expect that the pest increases and predator declines. Apparently, this is what has been observed in practice, and it suggests that indiscriminate use of man-made controls can sometimes worsen rather than ameliorate the natural balance.

An important stability tool is provided by Theorem 3.1. To find an appropriate Liapunov function, consider first the simplest kind of predator–prey system in which prey is allowed to grow exponentially instead of logistically. Since a predator P_2 will effectively limit the growth of P_1, one can argue that the prey can never grow too large anyway. With this simplification, 4.21 reduces to

$$\dot{N}_1 = rN_1 - \alpha N_1 N_2$$
$$\dot{N}_2 = -cN_2 + \beta N_1 N_2. \tag{4.30}$$

The nontrivial equilibrium of 4.30 occurs at $\overline{N}_1 = c/\beta$, $\overline{N}_2 = r/\alpha$. A linearized stability analysis reveals that the corresponding linear system has eigenvalues with zero real parts, and so its equilibrium state is neutrally stable. This, of course, conveys no information about the nonlinear system 4.30. However, there is another way to proceed. To simplify matters, let us first rescale 4.30 by letting $x_1 = (\alpha/r)N_1$ and $x_2 = (\beta/K)N_2$. Then it is easily seen that 4.30 is equivalent to

$$\begin{aligned} \dot{x}_1 &= rx_1(1 - x_2) \\ \dot{x}_2 &= -cx_2(1 - x_1). \end{aligned} \tag{4.31}$$

The effect of this variable change is to eliminate two of the constants in the model. This is often a convenient device and will be employed more frequently in subsequent chapters. If $\dot{x}_2 \neq 0$, then 4.31 leads to

$$\frac{dx_1}{dx_2} = \frac{\dot{x}_1}{\dot{x}_2} = \frac{-rx_1(1 - x_2)}{cx_2(1 - x_1)}, \tag{4.32}$$

which is a separable differential equation. It can be integrated to yield

$$c(x_1 - \ln x_1) + r(x_2 - \ln x_2) = \text{constant}. \tag{4.33}$$

Thus, relation 4.33 must hold along orbits of 4.31. This looks like a conservation of energy type of relation, and it suggests that one define a function

$$V(\mathbf{x}) = c(x_1 - \ln x_1) + r(x_2 - \ln x_2) - (c + r). \tag{4.34}$$

Since $\eta - \ln \eta \geq 1$ for any scalar $\eta > 0$, it follows that $V(x) \geq 0$ with zero only at the equilibrium $\bar{\mathbf{x}} = \begin{pmatrix} 1 \\ 1 \end{pmatrix}$. Moreover, because of 4.33, $\dot{V}(\mathbf{x}(t)) = 0$ along orbits of 4.31. Therefore, V is a Liapunov function in the state space U consisting of the positive quadrant. Because $\dot{V}(\mathbf{x}) = 0$ everywhere on U, we conclude that $\bar{\mathbf{x}}$ is stable, but not necessarily a.s. In fact, the orbits which satisfy 4.33 are a family of closed level curves about $\bar{\mathbf{x}}$, and so stability is indeed neutral (Exercise 4.5.8).

To construct a Liapunov function for the more general quadratic system 4.25, it is necessary to redefine V slightly. If $\overline{\mathbf{N}}$ is the nontrivial equilibrium, let

$$V(\mathbf{N}) = c_1\overline{N}_1\left[\frac{N_1}{\overline{N}_1} - \ln\left(\frac{N}{\overline{N}_1}\right) - 1\right] + c_2N_2\left[\frac{N_2}{\overline{N}_2} - \ln\left(\frac{N_2}{\overline{N}_2}\right) - 1\right]. \tag{4.35}$$

V is defined on the nonnegative quadrant and is nonzero except at $\overline{\mathbf{N}}$. The quantities c_1 and c_2 are positive constants to be determined. A calculation shows that along orbits of 4.25 one has

$$\begin{aligned}\dot{V}(\mathbf{N}(t)) &= c_1\left(N_1 - \overline{N}_1\right)\frac{\dot{N}_1}{N_1} + c_2\left(N_2 - \overline{N}_2\right)\frac{\dot{N}_2}{N_2} \\ &= c_1 a_1\left(N_1 - \overline{N}_1\right)^2 + c_1 b_1\left(N_1 - \overline{N}_1\right)\left(N_2 - \overline{N}_2\right) \\ &\quad + c_2 a_2\left(N_1 - \overline{N}_1\right)\left(N_2 - \overline{N}_2\right) + c_2 b_2\left(N_2 - \overline{N}_2\right)^2. \quad (4.36)\end{aligned}$$

Two special cases of 4.25 can now be treated. First the model 4.21 in which $a_1 = -r/K$, $a_2 = \beta$, $b_1 = -\alpha$, $b_2 = 0$. Let $c_1 = 1$, and choose c_2 to make all but the first term in 4.36 vanish. This is easily achieved by giving c_2 the value α/β. Then

$$\dot{V}(\mathbf{N}(t)) = -\frac{r}{K}\left(N_1 - \overline{N}_1\right)^2, \qquad (4.37)$$

which is negative except on the set R on which $N_1 = \overline{N}_1$. But $\dot{N}_1$ is nonzero except at $\overline{\mathbf{N}}$, and so the only invariant subset of R is $\overline{\mathbf{N}}$ itself. By Theorem 3.1, $\overline{\mathbf{N}}$ is globally a.s. on U.

In the model 4.22, a similar conclusion may be drawn when there is mild competition, as specified by the conditions in Figure 4.4d (Exercise 4.5.9).

Before leaving this section note that the simplified model 4.31 is not structurally stable (see the discussion in Section 3.2). All one has to do is add on a term $-(r/K)N_1$ to the first of these equations. This gives model 4.21 in which the qualitative nature of the orbits is completely different. Instead of neutrally stable cycles about $\overline{\mathbf{N}}$, we now have orbits that tend toward $\overline{\mathbf{N}}$. This is true no matter how small r/K is taken. That is, even the slightest perturbation of 4.31 completely changes the character of its solutions. In some sense, this is an unsatisfactory situation since two individuals who model the predator–prey system independently of each other could quite reasonably end up with either 4.21 or 4.30. We would want that small differences in these models be reflected in small changes in the orbit portraits, but this is not the case. Nonetheless there is a way out. In Chapter Six, a non-quadratic predator–prey model is introduced that appears to avoid this difficulty.

4.5 Exercises

4.5.1 Solve Equation 4.7 and interpret its solution.

4.5.2 Consider a growth model in which the growth rate $f(N)$ has the form indicated in Figure 4.3. What are the equilibria and which do you expect to be stable?

4.5.3 Do a linearized stability analysis of 4.14 about the zero rest state.

4.5.4 Interesting growth models are obtained from a study of chemical interactions. A simple case is that of two chemicals whose quantities at time t are $M_1(t)$ and $M_2(t)$. They combine to form a new chemical which is denoted by $M_3(t)$. Initially, we suppose M_3 to be zero. It is known that the rate of growth of M_3 is proportional to the product of M_1 and M_2. This means that if M_1, say, is held fixed, then the formation of M_3 increases in proportion to M_2 and vice-versa.

M_3 can also break apart into its constituents M_1 and M_2, and this occurs at a rate proportional to the level of M_3 itself. The roles of M_1 and M_2 are interchangeable, and so we might as well suppose that $M_1 = M_2$ for all time. A final assumption is that the rate of increase (decrease) of M_3 occurs at the same rate that M_1 decreases (increases). That is, $\dot{M}_3 = -\dot{M}_1$.

With this information, formulate a differential equation model for M_3.

4.5.5 Solve the equation in the previous exercise by separating variables. Interpret the solution.

4.5.6 In a predator–prey system, the predator P_1 has an alternate food source P_3 in addition to P_2. Write down a plausible model for the interaction between the three species. Assume that P_2 and P_3 do not interact among themselves.

4.5.7 Write down a reasonable model for the situation in which the two species P_2 and P_3 in the previous exercise compete for the same food source while they are also victims of predator P_1, which feeds on P_2 and P_3 at different rates.

4.5.8 Establish that the level curves defined by 4.33 are closed.

4.5.9 Using the Liapunov function 4.35, show that the nontrivial equilibrium depicted in Figure 4.4d is a.s.
Hint: 4.36 may be written as the quadratic form

$$\dot{V}(\mathbf{N}(t)) = \tfrac{1}{2}(\mathbf{N}(t) - \overline{\mathbf{N}})^{\mathrm{T}}(C\Lambda + \Lambda^{\mathrm{T}}C)(\mathbf{N}(t) - \overline{\mathbf{N}}),$$

where

$$\mathbf{N} = \begin{pmatrix} N_1 \\ N_2 \end{pmatrix}, \qquad C = \begin{pmatrix} c_1 & 0 \\ 0 & c_2 \end{pmatrix}, \qquad \text{and} \qquad \Lambda = \begin{pmatrix} a_1 & b_1 \\ a_2 & b_2 \end{pmatrix}.$$

It suffices to choose C so that $C\Lambda + \Lambda^{\mathrm{T}}C$ is negative-definite.

4.5.10 The predator–prey model can be modified to allow the prey to move into and out of its habitat at a constant rate σ. This leads to the following model:

$$\dot{N}_1 = rN_1 - \alpha N_1 N_2 - \sigma$$

$$\dot{N}_2 = -cN_2 + \beta N_1 N_2.$$

Deduce whether the nontrivial equilibrium is stable or not depending on the sign of σ (a positive σ indicates that the prey migrates out of its habitat or is harvested by some outside agent). Sketch the isoclines for this problem. Explain why it is possible for the prey to die out in finite time if $\sigma > 0$.

4.5.11 We are given a simplified model of conventional combat for two opposing forces in which there is no attrition due to self-inflicted losses:

$$\dot{N}_1 = -\alpha N_2$$

$$\dot{N}_2 = -\beta N_1.$$

Study this model by the methods developed so far in this book, and even obtain the explicit solution. Assume that N_1 and N_2 are initially positive quantities. Interpret the results you get.

4.5.12 An infectious disease spreads among a population by contact between individuals. $N_2(t)$ members of the population are infected at time t, and these either die or are quarantined away from the remaining individuals at a per-capita rate $\gamma > 0$. The remaining $N_1(t)$ members of the population are as yet uninfected but still susceptible to the disease by contact. Formulate a reasonable model for the interaction between susceptibles and infectives, study it by the methods so far developed, and interpret the results obtained.

4.5.13 An organic nutrient, such as phosphates or nitrates from detergents and fertilizers, enter a bay at a constant rate σ. Small floating plant organisms (algae) feed on the nutrient, and tiny swimming crustaceans graze on the algae. Because of the movement of the tides a fraction Q per unit time of the nutrients and algae are swept out to the ocean, never to return. Write down a reasonable differential equation model relating the levels of nutrient, algae, and crustaceans.

A Summary of Part 1

Before proceeding on to Part 2, let us highlight the main ideas and tools that have been introduced so far.

All the models arose from essentially two principles. One is *Newton's laws of motion* in the presence of forces that act on a body, and the other is the statement of *mass conservation*. These principles led to differential equation models whose solutions lie in a plane. Although it is possible to find explicit solutions in a few cases, there are many nonlinear equations that are not so accommodating. As a result, alternate approaches had to be introduced. Foremost among these is the notion of *stability or instability of an equilibrium state* from which it is often possible to infer something of the long-term behavior of the dynamical system.

The implementation of the stability concept relies on several tools. One of these is that of *linearization*. More powerful, but limited in applicability, are the use of *Liapunov functions* and, at a more intuitive level, the geometric idea of *isoclines*. When *conservation of energy* is valid, one can also apply *phase plane arguments*. Finally, we saw that if an equilibrium is unstable, then it may be possible to stabilize it by means of a *feedback control*.

To simplify our discussion, we often invoked the idea that only *generic properties* of a system need be considered since they represent the typical case. In connection with this is the concept of *structural stability* in which small changes in a model is sometimes reflected by small changes in orbits. Frequently used was the fact that if the dynamical equations are smooth, then *solutions are uniquely defined by their initial conditions* and hence do not cross.

We have also seen the pitfalls of using a continuous model when a discrete one is intended.

All these ideas will recur many times in subsequent chapters, and even though the models to be discussed later are more sophisticated, the core notions introduced so far are the key to understanding what is going on. Of course, other tools are to be added to the arsenal of methods a modeler needs to know, but a grasp of Part 1 paves the way.

Part 2

Further Thoughts and Extensions

Chapter Five

Motion in Time and Space

5.1 Conservation of Mass, II

Our models have so far been limited to physical situations in which time is the only independent variable. But there are interesting processes that depend on a spatial dimension as well. This chapter contains a sampling of models in which spatial considerations enter independently of time.

The added complication of space leads to partial rather than ordinary differential equations. The subject of partial differential equations is far reaching, and we cannot attempt an overview of its theory. Instead a single nonlinear equation is derived in this section, and various special cases are then taken up in the remainder of the chapter. In doing so, it will be found that just a few tools suffice to give some insight into the nature of the solutions. One of the reasons that these tools (characteristics, traveling waves, steady-state assumption) are effective is that they generally reduce the problem to a study of ordinary differential equations. As far as partial differential equations are concerned, this chapter is therefore self-contained. The curious reader may find indications on further readings in the References, where there is also a discussion of the background sources for the various models that are introduced.

For simplicity, all motion is taken to lie along a single coordinate direction, the x axis. In this chapter, x will denote a scalar variable and not a vector.

The derivation of our basic equation depends on the conservation of mass principle, which was introduced in Chapter One, but this time in a slightly more involved way to allow for space to enter in an explicit way.

A substance moves this way and that along some medium which is identified with the x axis (for example, heat along a rod, cars moving down a highway, pollutants down a river, or algae tossed about in a cross-section of the ocean surface). Let $\rho(x, t)$ be the density or concentration of the substance at x and at time t (in units of mass per unit distance) and $q(x, t)$ its rate of flow past x at time t (in units of mass per unit time). $q \geq 0$ is taken to mean that the net flow is from left to right. Both ρ and q are assumed to be smooth functions.

Conservation of mass says that in the absence of external additions or subtractions of mass, the rate at which the substance (or "stuff", as we often refer to it) changes inside an interval of length Δx within the medium must be due solely to the rate at which it moves across the boundary of that interval.

Since $\int_x^{x+\Delta x} \rho(s, t)\, ds$ is the total mass inside an interval of length Δx, this principle translates into the mathematical statement

$$\frac{d}{dt} \int_x^{x+\Delta x} \rho(s, t)\, ds = q(x, t) - q(x + \Delta x, t). \tag{5.1}$$

The right side of 5.1 represents the net flow into the left end of the interval at time t minus the net flow out at the right end. This difference is the net flow across the boundary of Δx at a fixed instant of time. If $q(x, t)$ were positive and $q(x + \Delta x, t)$ negative, for example, then all flow is into Δx and 5.1 would be positive, signifying that the mass inside the interval is increasing.

Now suppose that some of the stuff can move into or out of Δx from outside the given medium. This represents an additional source or a loss (sometimes called a "sink"), which must be accounted for (examples: the rod is not insulated against heat loss, the highway has entrances or exits, new pollutants run off into the river along its banks, algae reproduce themselves in the ocean). Let $k(x, t)$ be the rate at which the density of sources or sinks is changing. Assume that k is a smooth and known function. $k \geq 0$ is taken to mean a net source; otherwise it is a net sink.

In practice not all sources or sinks are smoothly distributed. Pollutants, for instance, may be discharged into a river at some specific sewage outfall and, for that matter, cars enter a highway at specific locations. These so-called point sources (or sinks) will be considered separately.

In any event, $\int_x^{x+\Delta x} k(s, t)\, ds$ is the rate at which the substance is moving into or out of Δx from sources or sinks external to the given medium. In order to maintain a balance of mass, this quantity must be added to the right side of 5.1. This then accounts for the rate at which the mass within Δx is changing due to outside influences in addition to effect of movement within the medium itself:

$$\frac{d}{dt}\int_x^{x+\Delta x} \rho(s, t)\, ds = q(x, t) - q(x + \Delta x, t) + \int_x^{x+\Delta x} k(s, t)\, ds.$$

Divide this expression by Δx. Since

$$\frac{d}{dt}\int_x^{x+\Delta x} \rho(s, t)\, ds = \int_x^{x+\Delta x} \frac{\partial \rho(s, t)}{\partial t}\, ds = \frac{\partial \rho(x, t)}{\partial t}\Delta x + o(\Delta x), \tag{5.2}$$

where $o(\Delta x)$ denotes terms which go to zero faster than Δx, then as $\Delta x \to 0$, we obtain in the limit

$$\frac{\partial \rho}{\partial t} = -\frac{\partial q}{\partial x} + k. \tag{5.3}$$

This is our basic partial differential equation. Note that the negative sign in 5.3 is reasonable. For suppose $\partial \rho/\partial t > 0$. This implies that stuff is bunching up at x at time t since its concentration is increasing. Therefore, the rate at which it flows past x should be decreasing or, to put it another way, $\partial q/\partial x$ must be negative. That is indeed what 5.3 shows.

Relation 5.3 involves two unknowns ρ and q, and so in order to be useful, we need to postulate various instances in which q can be expressed as some (smooth) function of ρ. If $q \equiv q(\rho)$ is given, then the equation becomes

$$\frac{\partial \rho}{\partial t} = -\frac{dq}{d\rho}\frac{\partial \rho}{\partial x} + k, \tag{5.4}$$

as is readily verified by the chain rule of differentiation. The equation is still too general, and some additional assumptions are needed to make it more tractable. Observe that $\dfrac{dq(\rho)}{d\rho}$ is typically a nonlinear function of ρ.

Two special cases are of interest to us. The first is a situation called *advection* in which $q = u\rho$, for some $u(x, t)$. The function u has units of distance per unit time and can be interpreted as the velocity of the stuff as it moves along the axis. One case of advection is particularly simple, namely when the stuff moves at a constant velocity c. Then $dq/d\rho$ is identical with velocity. This can also be seen in another way. In time Δt, the substance moves a distance $\Delta x = c\,\Delta t$, and so

$$\int_{x}^{x+c\,\Delta t} \rho(s, t)\, ds \tag{5.5}$$

represents all the stuff moving past x during Δt. Dividing by Δt and letting Δt shrink to zero in 5.5 must then give the rate at which the substance changes at x, namely $q(x, t)$. Therefore,

$$q(x, t) = \lim_{\Delta t \to 0} \frac{1}{\Delta t} \int_{x}^{x+c\,\Delta t} \rho(s, t)\, ds = c\rho(x, t). \tag{5.6}$$

A second way in which q is expressed as a function of ρ leads to what is called *diffusion*. We take our cue from heat flow, which is known to occur at a rate proportional to the temperature gradient (that is, proportional to the rate at which temperature, which is a measure of the density of heat, is changing with respect to position). Moreover, this flow is always from a higher to a lower temperature. This is expressed as

$$q(x, t) = -\nu \frac{\partial \rho(x, t)}{\partial x}, \tag{5.7}$$

with $\nu > 0$ a constant of proportionality. The minus sign shows that if $\partial\rho/\partial x > 0$ (so that temperature is increasing at x and at time t), then net flow is in the opposite direction (right to left). The movement of heat comes about from the random motion of colliding molecules, and so diffusion characterizes the movement, which is hither and thither along the x axis, the very opposite of advection. Other physical processes in which erratic motion is to be expected can be modeled by 5.7.

To summarize: under advection one has $q = \rho u$, whereas diffusion stipulates that $q = -\nu\,\partial\rho/\partial x$. By substituting the first of these relations into 5.4 we find that

$$\frac{\partial \rho}{\partial t} = -\frac{\partial(\rho u)}{\partial x} + k. \tag{5.8}$$

Substituting the diffusion relations into 5.4 gives

$$\frac{\partial \rho}{\partial t} = \nu \frac{\partial^2 \rho}{\partial x^2} + k. \tag{5.9}$$

Equation 5.9 in which $k = 0$ is known as the *heat equation* and is linear since ρ and its derivatives appear linearly.

More generally, if both advection and diffusion are present, then one can sum these effects and obtain

$$\frac{\partial \rho}{\partial t} = \nu \frac{\partial^2 \rho}{\partial x^2} - \frac{\partial(\rho u)}{\partial x} + k. \tag{5.10}$$

Most of this chapter consists of applications of Equation 5.10 in a variety of model settings.

5.2 Algae Blooms

An accumulation of nutrients favorable to the growth of certain tiny marine organisms called phytoplankton (or "plankton" for short) stimulates their rapid reproduction in the ocean. These plankton blooms appear as patches on the surface of the water. Outside a given patch the conditions for growth are unfavorable and so none takes place. The plankton are swept to and fro by wave action and the wind, and so their motion within an island-like patch may be considered random, much like the molecules in motion which lead to heat. This suggests that a diffusion model is appropriate. As they move about some are carried out past the patch boundaries and are lost. The only compensation for these losses is internal reproduction, which constantly adds to the population. This is considered to be a source term.

The tradeoff between reproduction and eventual loss at the boundary's edge means that there must be a minimal size for a patch if the population is not to die out. Diffusive motion in a small enough environment will outstrip the ability of remaining plankton to provide new recruits who can replace those lost by dispersion.

Although the ocean surface is two-dimensional, we assume that the plankton density is homogeneous in one of the two dimensions and varies only along the remaining x coordinate. This restriction does not seriously detract from the conclusions reached. Indeed, it is often true in the modeling of spatio-temporal processes that much of the interesting behavior is already apparent in a one-dimensional setting, giving results that are qualitatively similar to those obtained with more spatial coordinates.

In our case, then, the problem is to find the minimum length L of patch (in a one-dimensional slice of the ocean) that permits a plankton bloom to occur.

By analogy with the temporal model $\dot{N} = rN$ of rapid population growth (Equation 4.1), one may assume that plankton density increases at a rate proportional to its present density ρ. Since reproduction is considered a source term, this means that

$$k(x, t) = r\rho(x, t). \tag{5.11}$$

If logistic growth is a more realistic hypothesis, let

$$k(x, t) = r\rho(x, t)\left[1 - \frac{\rho(x, t)}{K}\right]. \tag{5.12}$$

Having now decided that a diffusive model is apt, the source assumptions combine with 5.9 to give either

$$\frac{\partial \rho}{\partial t} = \nu \frac{\partial^2 \rho}{\partial x^2} + r\rho \tag{5.13}$$

or

$$\frac{\partial \rho}{\partial t} = \nu \frac{\partial^2 \rho}{\partial x^2} + r\rho\left(1 - \frac{\rho}{K}\right). \tag{5.14}$$

For the moment, let us concentrate on the easier of the two, Equation 5.13. Since it is linear in ρ (whereas 5.14 is not), we can try a solution technique that is often used with linear partial differential equations. One simply tries to find a solution ρ that can be written as a product of terms, one involving x and the other t. This is called *separation of variables*:

$$\rho(x, t) = A(x)B(t). \tag{5.15}$$

Substituting 5.15 into 5.13, and then letting superscript dot denote time derivatives and superscript prime the space derivative gives, after dividing through by $\nu\rho = \nu AB$ and rearranging,

$$\frac{\dot{B}}{\nu B} - \frac{r}{\nu} = \frac{A''}{A}. \tag{5.16}$$

Now comes a crucial step. The two sides of 5.16 are patently quite different. One is a function of x, the other of t, each independent variables. How can they be equal? The answer, of course, is that this can only happen if they are both equal to an identical constant, say λ. Therefore, the original problem has now been reduced to solving two separate ordinary linear differential equations:

$$\begin{aligned} A'' - \lambda A &= 0 \\ \dot{B} - (\nu\lambda + r)B &= 0. \end{aligned} \tag{5.17}$$

The second of these equations is separable and immediately integrable:

$$B(t) = \text{const} \cdot e^{(\nu\lambda + r)t}. \tag{5.18}$$

We know from Chapter Four that the growth that a population sustains over time, when the net growth rate is r, is proportional to e^{rt}. If λ is positive, then 5.18 exceeds this growth rate, and this is untenable. It follows that λ is nonpositive. Let $\alpha = -\lambda$. Then

$$B(t) = \text{const} \cdot e^{(r - \alpha\nu)t}. \tag{5.19}$$

The solution to the first of 5.17 is the same as that of the harmonic oscillator of Chapter One and can be written as a linear combination of $\sin\sqrt{\alpha}\, x$ and $\cos\sqrt{\alpha}\, x$. It follows that for some constants c_1 and c_2, $\rho(x, t)$ is given as

$$\rho(x, t) = e^{(r - \alpha\nu)t}\left(c_1 \cos\sqrt{\alpha}\, x + c_2 \sin\sqrt{\alpha}\, x\right). \tag{5.20}$$

At this point, we model the effect of the patch boundaries by stipulating the boundary conditions

$$\rho(0, t) = \rho(L, t) = 0. \tag{5.21}$$

The patch is taken to be the interval from zero to L, and 5.21 says that population density at the edges must be zero. This models the unfavorable conditions outside this interval. Therefore,

$$\rho(0, t) = c_1 e^{(r-\alpha\nu)t} = 0,$$

which implies $c_1 = 0$ as the only possibility. Then

$$\rho(L, t) = c_2 e^{(r-\alpha\nu)t} \sin\sqrt{\alpha}\, L = 0.$$

Hence, $c_2 = 0$ is certainly a way of satisfying this relation, but 5.20 then gives ρ identically zero, an unacceptable solution. So it must be that $\sin\sqrt{\alpha}\, L = 0$. The smallest positive L that achieves this, other than $L = 0$, is $L = \pi/\sqrt{\alpha}$ and so $\alpha = (\pi/L)^2$.

Necessarily $r - \alpha\nu \geq 0$, otherwise we see from 5.20 that $\rho(x, t) \to 0$ as $t \to \infty$, which is again what we are trying to avoid. From the identity $\alpha = (\pi/L)^2$, it is therefore apparent that

$$r - \nu\left(\frac{\pi}{L}\right)^2 \geq 0$$

or

$$L \geq \pi\sqrt{\frac{\nu}{r}}\,. \tag{5.22}$$

This is the solution to our problem since any L smaller than its lower bound in 5.22 does not permit a sustained growth of plankton. Note that as the growth rate r increases, the critical size of L is allowed to get smaller, whereas if the tidal dispersion increases (as measured by ν, which indicates the intensity of diffusion), L must also increase. Both conclusions are consonant with what we expect.

The separation of variables idea worked because the equation was linear. In dealing with Equation 5.14, which also arises in other quite dissimilar models, a different approach is needed.

Let us imagine that the interchange between internal growth and loss at the boundary has been going on for a long time so that a population density is eventually reached that depends only on position and not on the time at which it occurs. In essence, the system has reached equilibrium with its environment, a situation called *steady state*. Under this condition of ρ no longer depending explicitly on time, it is a useful approximation to suppose that $\partial\rho/\partial t$ is zero. Partial derivatives with respect to x now effectively become ordinary derivatives in a single independent variable. Equation 5.14 then simplifies to

$$\frac{d^2\rho}{dx^2} + \frac{r\rho}{\nu}\left(1 - \frac{\rho}{K}\right) = 0. \tag{5.23}$$

This is a second-order equation of the form $p'' + f(p) = 0$, which was considered in Section 2.4. It satisfies a conservation of energy relation:

$$\frac{(\rho')^2}{2} + U(\rho) = C \tag{5.24}$$

in which $U(p)$ is a *potential function* and C a constant. In our case,

$$U(\rho) = \frac{r}{\nu}\int_o^\rho s\left(1 - \frac{s}{K}\right) ds = \frac{r}{\nu}\left(\frac{\rho^2}{2} - \frac{\rho^3}{3K}\right). \tag{5.25}$$

The function U has the form shown in Figure 5.1 in which the phase plane diagram is also drawn, corresponding to different values of the constant C. We suggest that you refer back to Examples 2.3, 2.4, and 2.5 of Chapter Two to understand how the phase diagram is constructed.

Note that if 5.23 is written as a first-order system, by letting $u_1 = \rho$ and $u_2 = \rho'$, then

$$\begin{aligned} u_1' &= u_2 \\ u_2' &= -\frac{ru_1}{\nu}\left(1 - \frac{u_1}{K}\right), \end{aligned} \tag{5.26}$$

which has equilibria at $\mathbf{u} = \begin{pmatrix} 0 \\ 0 \end{pmatrix}$ and $\mathbf{u} = \begin{pmatrix} K \\ 0 \end{pmatrix}$. A linearized stability analysis shows that the first equilibrium has neutral stability while the second is a saddle (Exercise 5.8.2). These assertions have their counterparts in Figure 5.1

In a steady state, the boundary conditions become $\rho(0) = \rho(L) = 0$. If a nontrivial solution to 5.23 exists, then it must begin and end on the

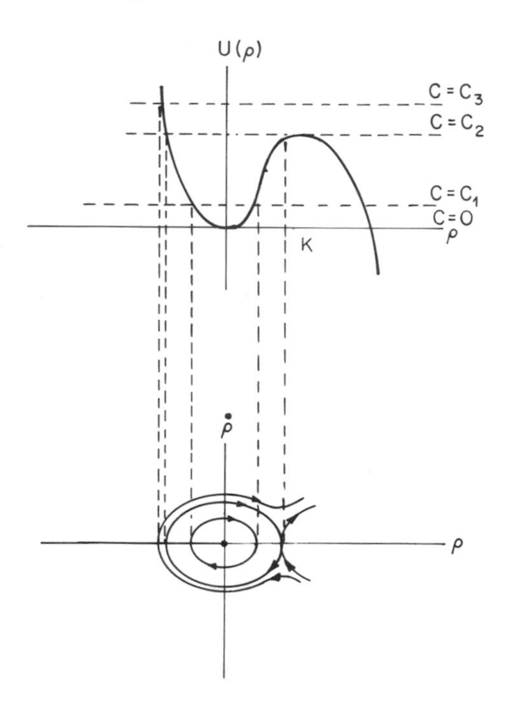

Figure 5.1 Phase portrait of Equation 5.23. Orbits are shown for different "energy levels" C. The points where ρ is zero or K on the horizontal axis are equilibrium states.

vertical ρ' axis shown in Figure 5.1. Since $\rho \geq 0$ for all x, this can only occur if the orbit begins on the positive ρ' axis and moves clockwise around an oval-shaped path. If the initial point is outside the region bounded by the separatrix (stable manifold) leading to the saddle $\begin{pmatrix} K \\ 0 \end{pmatrix}$, then such a path is impossible. It follows that $\rho < K$ for all x.

When the orbit begins on the separatrix itself, then since the saddle is an equilibrium state, it can only approach it asymptotically (the uniqueness principle again!). This means that the orbit corresponds to an interval of infinite length and in this case the boundary condition $\rho(L) = 0$ cannot be satisfied since the orbit is not able to continue on to the ρ' axis. An explicit expression for L is obtained by noting that from relation 5.24 one has

$$p'(x) = \pm\sqrt{2[C - U(\rho)]}\ .$$

Let an orbit cross the horizontal axis at some $\rho_1 < K$. Since the path is symmetrical about the ρ axis, consider only half of it as ρ goes from zero to ρ_1. It covers a distance $L/2$. Then, as in Exercise 2.5.6,

$$L = 2\int_0^{L/2} dx = 2\int_0^{\rho_1} \frac{d\rho}{\rho'} = \sqrt{2}\int_0^{\rho_1} \frac{d\rho}{\sqrt{C - U(\rho)}}. \tag{5.27}$$

We assume here that ρ' is not zero. Otherwise, ρ is a constant (namely zero, because of the boundary conditions), which is the trivial solution. Because of relation 5.24, one can evaluate the constant C as $U(\rho_1)$ since the derivative of ρ is zero at that point. Therefore,

$$L = \sqrt{2}\int_0^{\rho_1} \frac{d\rho}{\sqrt{U(\rho_1) - U(\rho)}}. \tag{5.28}$$

Figure 5.1 shows that $U(\rho)$ increases with ρ for ρ less than K. Hence, the term inside the square root in 5.28 is positive. Also the sign of this root is positive since as ρ goes from zero to ρ_1, the derivative $\rho'(x)$ remains nonnegative. It is also apparent from the figure that ρ_1 increases with C. That L also increases with C is less obvious.

Relation 5.28 shows that if a nontrivial steady state orbit is to exist on an interval of length L, then L must not be less than a quantity which is roughly $\pi\sqrt{\nu/r}$, which is the same qualitative conclusion reached before in the simpler model 5.13. To see this note that if ρ is small then $U(\rho)$ is roughly $r\rho^2/2\nu$, and so 5.28 is readily integrated to be $\pi\sqrt{\nu/r}$ (Exercise 5.8.4). The same conclusion can be ascertained from 5.23 since this represents an equation for harmonic motion with period $2\pi\sqrt{\nu/r}$ when ρ is small. Therefore, a half period has length $\pi\sqrt{\nu/r}$.

Figure 5.1 demonstrates that ρ increases to a maximum value at the point $x = L/2$ and then decreases again to zero at $x = L$.

5.3 Pollution in Rivers

An organic pollutant, such as human and animal fecal wastes, is thoroughly mixed in the water of a river which is moving downstream at a constant velocity c. The concentration ρ of the pollutant in the river is homogeneous in all directions except that of the downstream flow, which we take to be from left to right along the x axis. The river is thereby modeled by an advective flow in one dimension. Diffusive effects due to river turbulence and irregularities in its contours as it meanders downstream are all ignored. However, the pollutant is allowed to decay in the water due to bacterial action, which gradually decomposes it. This represents a sink term.

Let k be the rate at which the pollutant density is degraded. We assume it to be proportional to the density itself:

$$k(x, t) = -\mu\rho(x, t),$$

where μ is a proportionality constant that measures the efficiency of bacterial action. Since an advective model is appropriate, we adopt 5.8 in the special instance of $u = c$, a constant. The equation is then

$$\frac{\partial \rho}{\partial t} = -c\frac{\partial \rho}{\partial x} - \mu\rho. \tag{5.29}$$

This is a linear first-order partial differential equation, which can be solved by a simple device. First of all, let us reduce it to a form in which the sink term does not appear explicitly. This can be done by letting

$$v(x, t) = \rho(x, t)e^{\mu x/c}.$$

Substituting into 5.29 quickly gives

$$\frac{\partial v}{\partial t} = -c\frac{\partial v}{\partial x}. \tag{5.30}$$

Since 5.30 contains no sources or sinks one expects that however the density v is distributed initially on the river, this same distribution will persist over time except that it should be continuously translated to the right to account for the movement of the river. This leads us to surmise that $v(x, t)$ can be written in terms of a single variable

$$s = x - ct, \tag{5.31}$$

which is a translation of position x by an amount ct. Define a function $H(x, t)$ by

$$H(s, t) = v(s + ct, t) = v(x, t).$$

Then

$$\frac{dH}{dt} = c\frac{\partial v}{\partial x} + \frac{\partial v}{\partial t} = 0$$

by virtue of 5.30. Therefore, H does not depend explicitly on t:

$$H(s, t) = f(s),$$

for some (as yet unknown) function f. It follows that

$$v(x, 0) = f(x). \tag{5.32}$$

This shows that f is in fact the initial distribution of the density v. Denote this known initial value by $v_0(x)$. Since 5.32 is true for all possible real numbers x including the value s, the solution is given as

$$v(x, t) = f(s) = v_0(x - ct). \tag{5.33}$$

This is called a *traveling wave* solution to 5.30 because the initial density distribution is propagated downstream as a wave front. Figure 5.2 gives two pictorial representations of this. Traveling wave solutions are to be encountered again in this chapter.

Since v is constant for any fixed value of $x - ct$, it has the same value for each x, t that lie on the straight line $x - ct = s$, a different value for each choice of x. There is therefore a family of straight lines, called *characteristics*, along each of which v is constant. Figure 5.3 shows the characteristics as lines having slope c in the x, t plane. They are the level

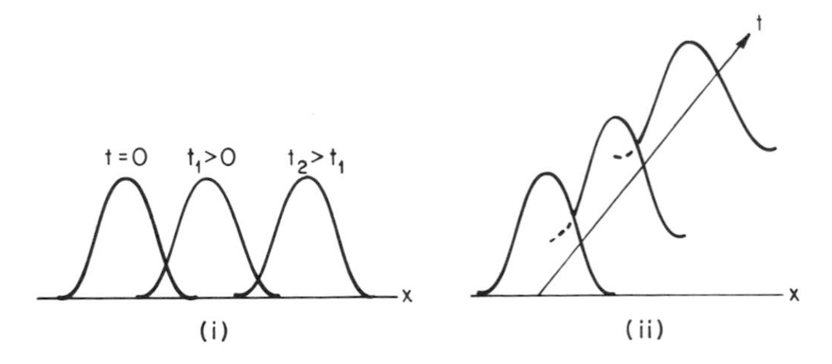

Figure 5.2 Two representations of the traveling wave solution of Equation 5.30. Case (i) shows how the initial density propagates downstream over time while (ii) displays the same thing on the x, t plane.

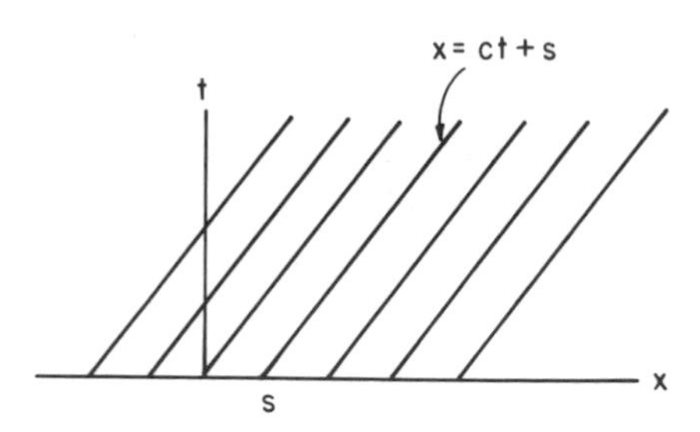

Figure 5.3 Characteristics corresponding to the traveling wave solution of Equation 5.30. The straight lines of slope c in the x, t plane are level curves of the density distribution in Figure 5.2ii.

curves of the surface shown in Figure 5.2ii. Once the characteristics are known, we have an explicit solution to an equation for all x, t provided the value of v is known at $t = 0$. It maintains the same value along the entire characteristic emanating from the x axis at the initial time. This will prove to be a powerful idea in Section 5.4.

Returning now to the original Equation 5.29, assume that $\rho(x,0) = \rho_0(x)$ is given. Then, since $v(x, t) = \rho(x, t)\, e^{\mu x/c}$, it follows that $v_0(x) = \rho_0(x)\, e^{\mu x/c}$, and therefore

$$\rho(x, t) = \rho_0(x - ct)e^{-\mu t},$$

in which the traveling wave is damped as t increases to account for the decay of pollutant over time.

Let us now modify the model to account for the practical situation in which there is a source of pollution at some given point on the river, which we take to be $x = 0$. A typical source is the discharge of a sewage treatment plant having an outfall on the river. Before the plant begins operating at $t = 0$, the river is clean. We want to determine the density of pollution downstream at any future time and location.

Pollutants are added to the river at a rate γ by the dumping. Define U by

$$U(t) = \begin{cases} 1 & \text{if } t \geq 0 \\ 0 & \text{if } t < 0. \end{cases}$$

Then we can write down a boundary condition:

$$\rho(0, t) = \gamma U(t). \tag{5.34}$$

This is the way in which a point source is introduced into our model. Since this source exists at only one location, we do not have a smooth source term to work with. Using a boundary condition gets us around this difficulty. However, in order to find $\rho(x, t)$ a slightly altered traveling wave solution must be sought. This is because we do not have an initial density distribution available. Instead, 5.34 provides a density distribution for all time at a single point. So this time define a variable η by

$$\eta = t - \frac{x}{c},$$

and a function

$$H(x, \eta) = v\left(x, \eta + \frac{x}{c}\right) = v(x, t).$$

Proceeding, then, as we did above shows (Exercise 5.8.6) that the traveling wave solution to 5.30 is

$$v(x, t) = \gamma U\left(t - \frac{x}{c}\right), \tag{5.35}$$

and so

$$\rho(x, t) = \gamma U\left(t - \frac{x}{c}\right) e^{-\mu x/c}. \tag{5.36}$$

Observe that $\rho(0, t) = \gamma U(t)$, as required.

The solution 5.36 shows that if $x > ct$, then $\rho(x, t) = 0$, which is reasonable since it takes a time t for the pollutant to reach a downstream location $x = ct$ from the outfall. The exponential factor indicates that the pollutant density decreases as the substance is decomposed by bacteria during its journey down the river (Figure 5.4).

Now let us consider a more interesting model. The bacterial decomposition of the pollutant requires an uptake of dissolved oxygen in the water. As the pollutant is degraded, oxygen is used up. Let $\delta(x, t)$ be the density of dissolved oxygen in the river. Its maximum value, which depends on temperature, is δ_m. We assume it to be a known fixed quantity.

The rate at which the oxygen dissipates is the same as that of pollutant decay and is proportional to the pollutant concentration. This

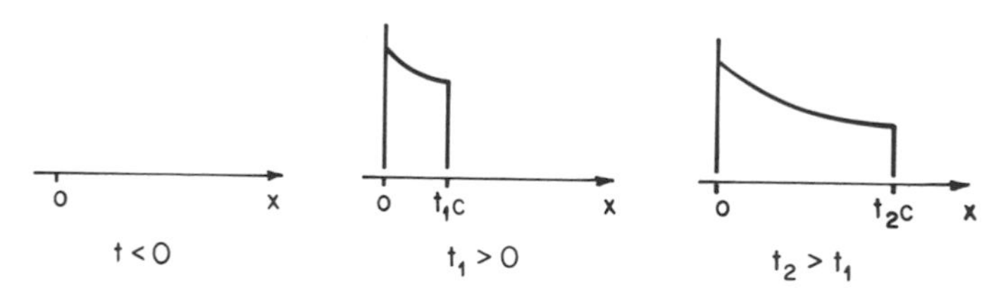

Figure 5.4 Pollutant density obtained from 5.36 and shown for three different times. Since the river is unpolluted upstream from the source at $x = 0$, we define $\rho(x, t)$ to be zero for $x < 0$.

is a sink term given as

$$k_1(x, t) = -\mu\rho(x, t).$$

There is also a source term due to the fact that the river surface, in contact with air above, draws oxygen in from the atmosphere by a process known as re-oxygenation. This happens at a rate proportional to the difference between the saturation level δ_m and the actual δ. Thus,

$$k_2(x, t) = \mu_1[\delta_m - \delta(x, t)].$$

Otherwise, the flow of oxygen is like that of the pollutant, and the advective model 5.8 is again the one to use with, however, the addition of a source and sink term:

$$\frac{\partial\delta}{\partial t} = -c\frac{\partial\delta}{\partial x} - \mu\rho + \mu_1(\delta_m - \delta). \tag{5.37}$$

We assume that $\mu_1 > \mu$ to ensure that self-purification of the river can overcome the degenerative effects of pollution.

The significance of dissolved oxygen is that it is sometimes used by sanitary engineers and marine scientists as a measure of the health of the water body. Oxygen poor waters are unhealthy for marine life and human consumption. It is therefore of considerable interest to correlate oxygen levels with the amount of pollutants that are present. Equation 5.37 expresses this relationship since it depends on ρ. To obtain ρ, one must also know the solution to 5.29, and so we have a pair of coupled partial differential equations to consider simultaneously. There are two boundary conditions, one for ρ (relation 5.34) and another for δ:

$$\delta(0, t) = \delta_m U(t). \tag{5.38}$$

This last condition expresses the fact that the oxygen level is at its highest value right at the source since the upstream water arrives untainted at $x = 0$. It begins to change as the pollutant moves downstream from there.

A plausible assumption is that the process of oxygen decay has reached a steady state after a sufficient lapse of time so that the pair of equations are unaffected by time, but depend only on position along the river. It is left to the reader to determine δ under this assumption. This leads to what is known as an *oxygen-sag curve* for a reason which is apparent after the solution is found (Exercise 5.8.7).

5.4 Highway Traffic

Cars move down a long single-lane highway from left to right. Because they have a finite size the density of cars at a given point is not well defined. However, by viewing the traffic flow from a far distance their size becomes negligible and it becomes plausible to approximate the density as a smooth function ρ.

It appears that an advective model is called for, as in the previous section. However, there is a notable difference in that velocity is not assumed to be constant. The cars should be able to move at different speeds if the model is to give interesting results. Also, there will be no smooth sources or sinks. Cars are neither created or destroyed on our road except through entrances and exits, which are point sources and sinks. These can be modeled by boundary conditions, much as in the water pollution model, and are not considered further.

From observations of actual traffic patterns, it is known that car velocity depends on traffic density in a roughly linear way. Suppose that u_m is the maximum speed any car is capable of on the highway (determined, perhaps, by speed limits and weather conditions) and ρ_m is the maximum possible density at which there is bumper to bumper traffic. Then u is related to ρ by

$$u(\rho) = u_m\left(1 - \frac{\rho}{\rho_m}\right). \tag{5.39}$$

This relation implies that each driver moves at the same velocity at any given density. As density increases, the spacing between cars diminishes, and so velocity changes with the distance between cars. This much is reasonable. As density changes the cars respond immediately. There is no acceleration or deceleration. Also fast cars cannot pass slow ones, and we ignore the effect of curves in the road, the presence of police cars, and go-slow zones. These assumptions are perhaps less reasonable, but the use of 5.39 gives insightful results nonetheless.

Because we have an advective model, $q = \rho u$ and Equation 5.8 give, after applying the chain rule of differentiation,

$$\frac{\partial \rho}{\partial t} = -\frac{\partial(\rho u)}{\partial t} = -\frac{dq}{d\rho}\frac{\partial \rho}{\partial x}. \tag{5.40}$$

From 5.39, one obtains

$$\frac{dq(\rho)}{d\rho} = u_{\mathrm{m}}\left(1 - \frac{2\rho}{\rho_{\mathrm{m}}}\right). \tag{5.41}$$

Since ρ appears nonlinearly in 5.40, the traffic model involves what is known as a first-order quasilinear partial differential equation. In the not very interesting case in which car velocity is constant, $u \equiv c$, 5.40 becomes identical to the river pollution equation 5.30, and the same kind of traveling wave solution applies. Let $\bar{\rho}$ be the average car density on the highway, and assume that the actual ρ varies but little from $\bar{\rho}$. Then Equation 5.40 can be linearized about $\bar{\rho}$, but this again leads to the same linear equation of river pollution (Exercise 5.8.8). We must instead treat the nonlinear version head-on.

An observer is initially at some location x_0 on the highway. He sees a traffic density $\rho_0 = \rho(x_0, 0)$. Suppose the observer moves at some velocity $\dot{x}$ that allows him to see a constant density. That is, he moves in such a way that if $x(t)$ is his position at time t, then

$$\rho(x(t), t) = \rho(x_0, 0) = \rho_0,$$

for all $t \geq 0$. This means that ρ has a constant value along some curve in the x, t plane whose components are $(x(t), t)$ and which emanates from $(x_0, 0)$. This curve is called a *characteristic* (see also the discussion in Section 5.2). By the chain rule of calculus,

$$\frac{d\rho(x(t), t)}{dt} = \frac{dx(t)}{dt}\frac{\partial \rho(x(t), t)}{\partial x} + \frac{\partial \rho(x(t), t)}{\partial t} = 0.$$

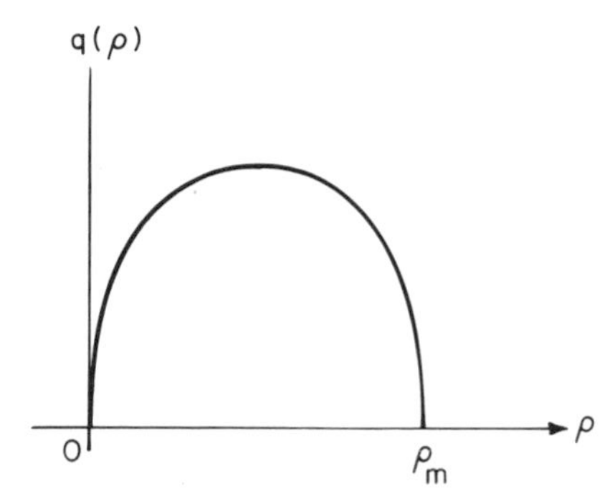

Figure 5.5 Traffic flow $q(\rho) = \rho u(\rho)$, based on relation 5.39. Here $q(\rho) = u_m \rho(1 - \rho/\rho_m)$.

Comparing this relation to Equation 5.40 shows that

$$\dot{x} = \frac{dq(\rho_0)}{\partial \rho} \tag{5.42}$$

along the characteristic on which $\rho = \rho_0$. Integrating 5.42 gives

$$x(t) = \frac{dq(\rho_0)}{d\rho} t + x_0, \tag{5.43}$$

which is a straight line having slope $\frac{dq(\rho_0)}{d\rho}$ (called the *local wave velocity*). Different values of $x_0 = x(0)$ give rise to generally different values of ρ_0, and so the slope may also vary with x_0. There is, therefore, a family of straight line characteristics in the x, t plane on each of which ρ remains constant. A knowledge of $\rho(x, 0)$ for any x effectively determines $\rho(x, t)$ for all time. Indeed, along a characteristic one obtains

$$\rho(x, t) = \rho(x_0, 0) = \rho\left(x - \frac{dq}{dt} t, 0\right) = \rho_0(s) \tag{5.44}$$

in which $s = x - \frac{dq}{d\rho} t$. This resembles a traveling wave solution except the wave velocity $\frac{dq(\rho_0)}{d\rho}$ is local; it can vary with ρ_0. Note that the local wave velocity is not the same as car velocity. In fact,

$$\frac{dq}{d\rho} = u + \rho \frac{du}{d\rho} \leq u$$

since u, being a decreasing function of ρ, has a negative derivative. Moreover, as one sees from Figure 5.5, $dq/d\rho$ can actually be negative if it is evaluated at any density that exceeds $\rho_m/2$. The characteristics can therefore have positive or negative slopes.

Two specific examples will illustrate these ideas. The second of these introduces a new complication since in this case the characteristics intersect. The resolution of this problem serves to introduce an important concept, that of *shock waves*.

Imagine what a traffic pattern looks like a few moments after a red light, behind which is a long column of waiting cars, turns green. Suppose the stop light, located at $x = 0$, had been on a long time. The moment it turns there are no longer any cars in front of it because they have all moved on long ago. Behind the light there is effectively an infinite line of bumper to bumper traffic. Then traffic begins to move. The first few cars on line go forward to the right of the green light, but most of them continue to be at a standstill. The light change occurs at $t = 0$. At this initial moment, the traffic density can be approximated by

$$\rho_0(x) = \rho(x,0) = \begin{cases} \rho_m, & x < 0, \\ 0, & x > 0. \end{cases} \tag{5.45}$$

From 5.41, we have

$$\frac{dq(\rho_0)}{d\rho} = \begin{cases} u_m, & \rho_0 = 0, \\ -u_m, & \rho_0 = \rho_m, \end{cases} \tag{5.46}$$

and so, from 5.34, the characteristics are obtained as straight lines defined by

$$x = \begin{cases} u_m t + x_0, & x_0 > 0, \\ -u_m t + x_0, & x_0 < 0. \end{cases} \tag{5.47}$$

These are shown in Figure 5.6. In the region far to the right of the origin where the characteristics have positive slope, the density is zero,

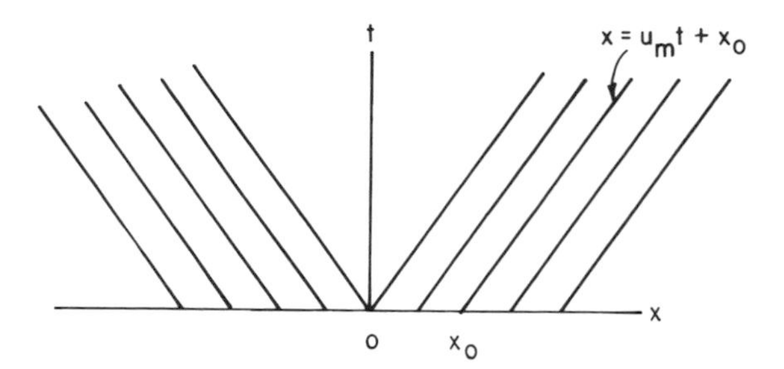

Figure 5.6 Characteristics defined by 5.47: straight lines with slope u_{m} for $x > 0$ and slope $-u_{\mathrm{m}}$ for $x < 0$.

which means that for any time $t > 0$ and position $x > 0$ in this region the traffic has not yet arrived after the light change. For x, t in the area to the left where the characteristics have negative slope, traffic continues to be at a standstill (density is at the maximum ρ_{m}) since the cars in front have not yet begun to move after the switch to a green light. What about the cone-shaped region in between? Here we expect traffic density to be decreasing from ρ_{m} as the first cars begin to move out at $t > 0$. The very first car can move at a maximum velocity u_{m} since the density in front of it is zero. The successive cars start up more slowly so there is a gradual thinning out of traffic. This may be seen more clearly by taking x_0 to be zero in 5.43.

Solving for ρ in this relation gives

$$\rho = \frac{1}{2}\rho_{\mathrm{m}}\left(1 - \frac{x}{u_{\mathrm{m}}t}\right). \tag{5.48}$$

By fixing t at some positive value, it is readily seen from this equation that ρ decreases linearly from ρ_{m} to zero as x goes from $-u_{\mathrm{m}}t$ to $u_{\mathrm{m}}t$ (Figure 5.7). We are looking for a continuous solution to the traffic equation, one which joins the solutions $\rho = \rho_{\mathrm{m}}$ for $x < 0$ to $\rho = 0$ for $x > 0$. The expression 5.48 certainly does this if one lets $dq/d\rho = x/t$, as the reader may quickly verify. In fact, this choice for $dq/d\rho$ is identical to setting x_0 to zero in 5.43.

The distance $-u_{\mathrm{m}}t$ behind the light is where the next car in line begins to move at time t, while $u_{\mathrm{m}}t$ is the distance in front of the light where the first car has just arrived (since it travels at maximum velocity). From this we see that the signal for the next car on line to begin moving

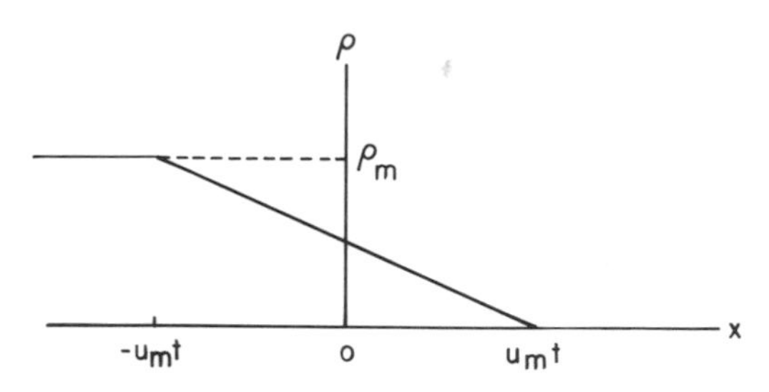

Figure 5.7 Traffic density at some time $t > 0$ after the red light turns to green.

is propagated backward at a velocity u_m. As ρ decreases, the characteristics are obtained from 5.43 with $x_0 = 0$, with slopes going from $-u_m$ to u_m. This gives a fan-like family of lines (Figure 5.8) which fill in the previously empty cone in Figure 5.6

The speed of individual cars in this fan-shaped zone is, of course, given by

$$u = u_m\left(1 - \frac{\rho}{\rho_m}\right). \tag{5.49}$$

Let $x(t)$ now denote the car's location at time t if it begins to move at position x_0 behind the light (at positive time $t_0 = -x_0/u_m$, as we see in Figure 5.6). Then $\dot{x} = u$ with $x = x_0$ at $t = t_0$. Combining 5.49 with 5.48 gives a differential equation:

$$\dot{x} = \frac{u_m}{2} + \frac{x}{2t}. \tag{5.50}$$

Solving for x in this equation tells us where the car is after it starts up. In particular, we can find out how long it takes for the car to pass the light. The result may be surprising (Exercise 5.8.9).

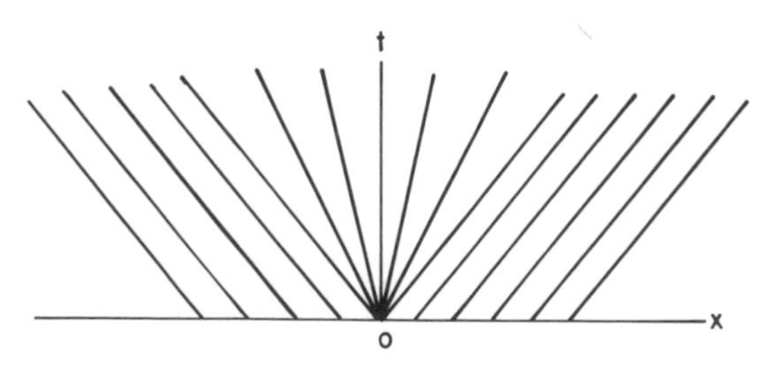

Figure 5.8 Characteristics for the traffic light problem.

Now consider the opposite situation in which the traffic density, instead of thinning out, is increasing with x. This means that the traffic is slower ahead and faster behind, and so cars in the rear will eventually catch up with those in front. This leads to an abrupt and catastrophic change in density at some point as illustrated in Figure 5.9.

The problem is seen mathematically if one draws the characteristics. To be specific let us approximate the density at the moment at which traffic of density $\rho_m/4$ is just about to catch up with standstill traffic of density ρ_m. What the model predicts in this case is a collision as the rear cars overtake the column of immobile cars in front. In fact, since the cars go immediately from a region of low density into a higher one, there is instant deceleration, or a crash!

Initially, then, we are given

$$\rho_0(x) = \rho(x,0) = \begin{cases} \dfrac{\rho_m}{4}, & x > 0, \\ \rho_m, & x < 0. \end{cases} \tag{5.51}$$

From 5.41 one finds

$$\frac{dq(\rho_0)}{d\rho} = \begin{cases} \dfrac{u_m}{2}, & \rho = \dfrac{\rho_m}{4}, \\ -u_m, & \rho = \rho_m. \end{cases}$$

The characteristics are therefore obtained from 5.43 as

$$x = \begin{cases} \dfrac{u_m t}{2} + x_0, & x_0 < 0, \\ -u_m t + x_0, & x_0 > 0. \end{cases}$$

These are sketched in Figure 5.10.

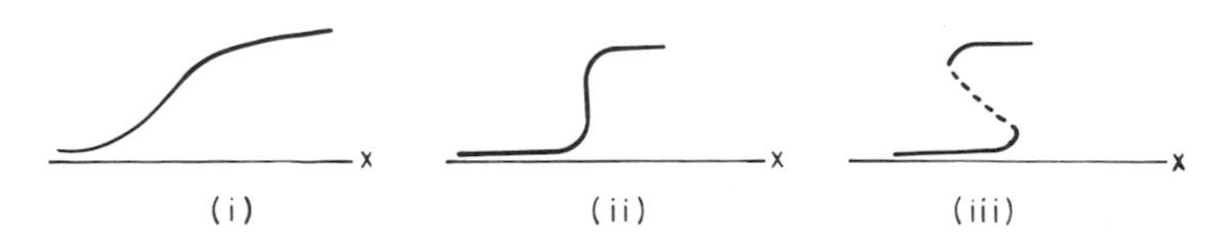

Figure 5.9 Evolution of traffic density ρ over time when the initial density has the form shown in (i). In (ii) the faster traffic has caught up with the slower. Finally, in (iii), the fast traffic passes the slow giving rise to a multivalued density. This forces us to rethink the model formulation.

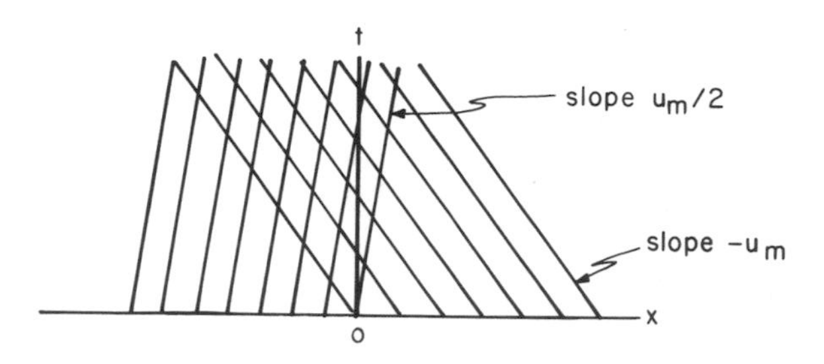

Figure 5.10 Characteristics for the traffic problem in which the initial density is given by 5.51.

One cannot help noticing that these lines intersect in a wedge-shaped region, which implies that in this zone ρ has two different values, one on each line, an unacceptable lack of uniqueness. Although in the previous traffic example, we were able to obtain a continuous solution ρ even though the initial data $\rho_0(x)$ was discontinuous, we must now face up boldly to the idea that this may not always be possible. In order to find a way out of the difficulty of non-uniqueness, we drop the requirement that ρ be everywhere smooth and permit it to be only piecewise continuous. This in turn necessitates that the relation $q = \rho u$ be re-interpreted.

To see what this entails, suppose x_s is a point on the highway where ρ has a jump discontinuity. In our problem, this occurs initially at $x_s = 0$. Now the point x_s may move as time goes on. Suppose, in fact, that x_s is between x and $x + \Delta x$ at some time t. Returning to the basic conservation of mass expression 5.1, we see the difficulty: the derivative cannot be taken inside the integral because of the jump in ρ. So rewrite the left side of 5.1 as

$$\frac{d}{dt}\int_x^{x+\Delta x} \rho(s,t)\,ds = \frac{d}{dt}\int_x^{x_s} \rho(s,t)\,ds + \frac{d}{dt}\int_{x_s}^{x+\Delta x} \rho(s,t)\,ds. \tag{5.52}$$

The density is smooth except for the jump at x_s. Let the left and right hand limits of ρ at this point be denoted as ρ_- and ρ_+ respectively. The derivative of 5.52 can be obtained by utilizing the following well-known rule for differentiating integrals whose limits of integration are variable:

if $\alpha(t)$, $\beta(t)$, $f(x, t)$ are given smooth functions, then

$$\frac{d}{dt}\int_{\alpha(t)}^{\beta(t)} f(x, t)\, dx = \int_{\alpha(t)}^{\beta(t)} \frac{\partial f(x, t)}{\partial t}\, dx + \dot{\beta} f(\beta(t), t) - \dot{\alpha} f(\alpha(t), t). \tag{5.53}$$

Applying this rule to 5.52 gives

$$\frac{d}{dt}\int_{x}^{x+\Delta x} \rho(s, t)\, ds = \dot{x}_s \rho_- + \int_{x}^{x_s(t)} \frac{\partial \rho(s, t)}{\partial t}\, ds$$

$$-\dot{x}_s \rho_+ + \int_{x_s(t)}^{x+\Delta x} \frac{\partial \rho(s, t)}{\partial t}\, ds. \tag{5.54}$$

Let $x \uparrow x_s$ and $x + \Delta x \downarrow x_s$. Then the integrals in 5.54 vanish. Combining this with 5.1 gives, with a corresponding interpretation for q_- and q_+ (or u_- and u_+):

$$\dot{x}_s = \frac{(q_+ - q_-)}{(\rho_+ - \rho_-)} = \frac{(\rho_+ u_+ - \rho_- u_-)}{(\rho_+ - \rho_-)}. \tag{5.55}$$

This differential equation can easily be solved since the right side is constant. In our example $\rho_+ = \rho_m$, $\rho_- = \rho_m/4$, and from 5.39 $u_+ = 0$, $u_- = 3u_m/4$. Therefore,

$$\dot{x}_s = -\frac{u_m}{4},$$

and since $x_s(0) = 0$,

$$x_s(t) = -\frac{u_m t}{4}. \tag{5.56}$$

Using understandably graphic language, x_s is the location of a shock and 5.55 is the characteristic of *shock wave*. The equation for x_s tells us

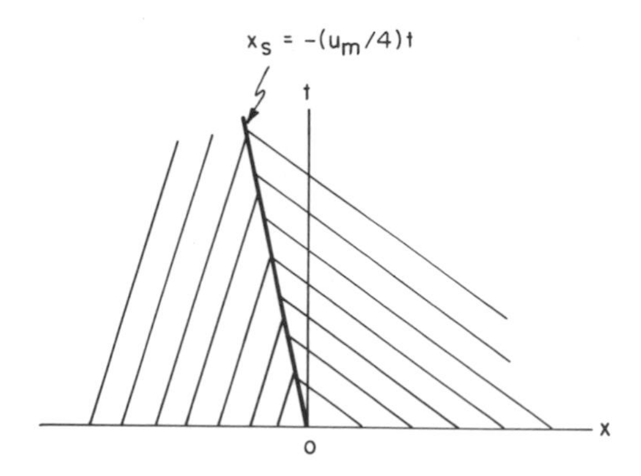

Figure 5.11 A shock wave separates the characteristics of slow and fast traffic on the highway.

where the discontinuity in traffic density is located at any time t. To the left and right of this line there are two different families of smooth characteristics. In effect, the shock wave characteristic is the interface between moving cars in one region and immobile ones in the other (Figure 5.11). Since the characteristic at x_s has a negative slope (in general it can be positive or negative), this means that the shock propagates backward through the traffic. Physically, this expresses the fact that the cars in the rear are catching up with (and crashing into) the stationary ones at points located increasingly to the left of the origin (Figure 5.12). Have you ever noticed how brake lights of the cars ahead seem to be moving backwards towards your car when there is sudden congestion at the forward location? This is a manifestation of the shock wave phenomenon.

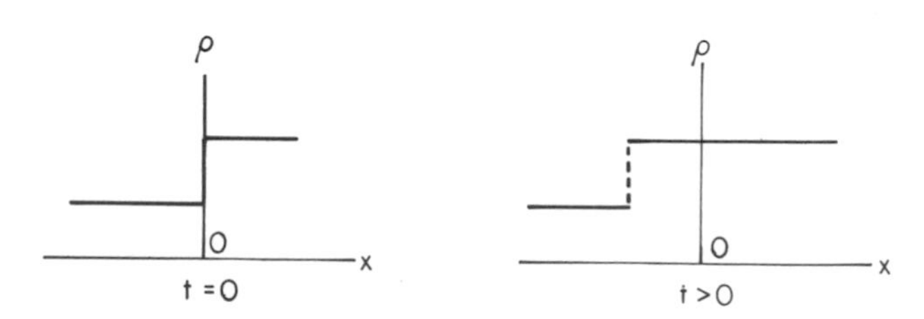

Figure 5.12 Standstill traffic density propagates backward when the initial density is given by 5.51.

An alternate approach to modeling congestion in the road ahead is to allow the drivers of cars to adjust their velocity not only according to the spacing between cars (viz, the density), but also to the rate at which this spacing is changing. The drivers slow down to compensate for an increasing density ahead. In effect we make u a function of both ρ and $\partial\rho/\partial x$.

A plausible model is obtained by replacing $q = \rho u$ with

$$\hat{q} = q - \nu \frac{\partial \rho}{\partial x},$$

or, equivalently,

$$\hat{u} = \frac{\hat{q}}{\rho} = u - \frac{\nu}{\rho}\frac{\partial \rho}{\partial x}, \tag{5.57}$$

where ν is a positive scalar. $\partial\rho/\partial x$ is positive if ρ is increasing, and so it is not surprising that $\hat{u}$ is less than the value u it would have if the change in density were ignored. As a result of using the modified velocity $\hat{u}$ in place of u, the traffic equation now becomes (Exercise 5.8.12):

$$\frac{\partial \rho}{\partial t} + u_{\mathrm{m}}\left(1 - \frac{2\rho}{\rho_{\mathrm{m}}}\right)\frac{\partial \rho}{\partial x} = \nu \frac{\partial^2 \rho}{\partial x^2}. \tag{5.58}$$

This equation, known as *Burger's equation*, seems to combine both advection and diffusion, and we suspect a traveling wave solution in which fast traffic never quite catches up with the slow but where an increasing density distribution propagates forward along the highway. Intuitively, this expresses the idea that cars are slowing down in response to congestion ahead while the whole column of traffic continues to move forward. Verification of the conjecture is left as Exercise 5.8.13.

5.5 A Digression on Traveling Waves

Having met some examples of traveling wave solutions, let us pause to note that this type of solution may be anticipated on the grounds of physical intuition in other spatio-temporal problems, such as the propagation of epidemics or the transmission of neuron pulses. To better grasp this important idea, we return to Equation 5.14. To simplify matters let the diffusion constant be unity and rescale ρ by writing $\rho = K\psi$. Then the equation becomes

$$\frac{\partial \psi}{\partial t} = \frac{\partial^2 \psi}{\partial x^2} + r\psi(1 - \psi). \tag{5.59}$$

We now assume an infinite spatial medium with no boundary conditions imposed. Relation 5.59 models the interaction between local population growth and global dispersion and is known as *Fisher's equation*.

In Figure 5.13 an intuitive argument is presented to support the idea that one might expect a traveling wave solution to 5.59, at least in the case in which the initial distribution $\psi(x, 0)$ of the density ψ is a step function.

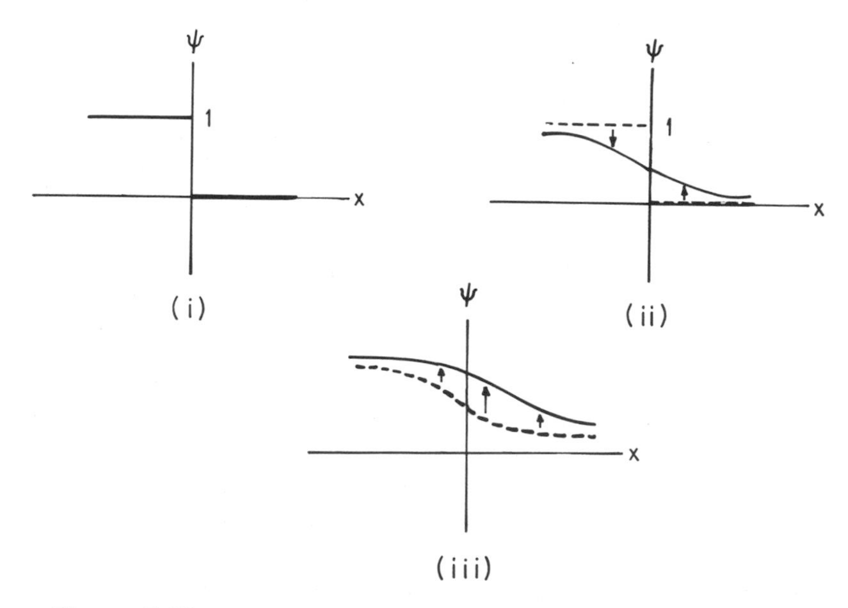

Figure 5.13 Intuitive argument for the temporal evolution of a solution of Equation 5.59 when the initial density is a step function. This is shown in a set of "snapshots" in space for different times. The step function is at (i). A bit later the step function (now dotted) has evolved to the solid line shown in (ii). What has happened is that diffusion has robbed from the dense population and added to the sparse one. Finally the wave form in (ii), now dotted, has evolved further (iii). This is because of intrinsic population growth, which is largest when ψ equals $\frac{1}{2}$ and is smallest near the endpoints zero and one. By repeating the actions of diffusion and growth in this manner, the original profile is seen to move to the right.

To examine this further, suppose that $\psi(x, t) = \phi(s)$ where $s = x - ct$ and $c > 0$ is some constant wave speed. Substituting this into 5.59 gives rise to an ordinary differential equation in ϕ:

$$-c\phi' = \phi'' + r\phi(1 - \phi). \tag{5.60}$$

In accordance with Figure 5.13, we wish to establish that 5.60 has a solution in which $\phi(-\infty) = 1$, $\phi(\infty) = 0$, and $\phi' < 0$. Also, since ϕ represents a population density whose maximum is one, it is entirely reasonable to require $0 \leq \phi \leq 1$. The function ϕ has the form of a stationary wave front. Conversely, having found such a solution ϕ to 5.60, it suffices to define $\psi(x, t)$ as $\phi(x - ct)$ to obtain a traveling wave solution to 5.59.

We write 5.60 as a first-order system by letting $u_1 = \phi$ and $u_2 = \phi'$ to obtain

$$\begin{aligned} u_1' &= u_2 \\ u_2' &= -ru_1(1 - u_1) - cu_2. \end{aligned} \tag{5.61}$$

There are two equilibria, $\begin{pmatrix} 0 \\ 0 \end{pmatrix}$ and $\begin{pmatrix} 1 \\ 0 \end{pmatrix}$, and the Jacobian matrix of the linearized system has a trace of $-c$ and a determinant of $r(1 - 2u_1)$. Therefore, $\begin{pmatrix} 1 \\ 0 \end{pmatrix}$ is a saddle point, and a straightforward computation further shows that there is an eigenvector corresponding to the positive eigenvalue that points into the negative quadrant at $\begin{pmatrix} 1 \\ 0 \end{pmatrix}$. Hence, a branch of the unstable manifold of the saddle can conceivably join up with $\begin{pmatrix} 0 \\ 0 \end{pmatrix}$ since the origin is stable. If such an orbit can indeed be found in the u_1, u_2 (ϕ, ϕ') phase plane, it means that there is a $\phi(s)$ with negative slope that satisfies $\phi(-\infty) = 1$, $\phi(\infty) = 0$. We will complete the argument for its existence later in Chapter 6 (Example 6.3).

In the meanwhile, observe that for a traveling wave to exist there must also be a minimal speed c. The reason for this is that the eigenvalues of the Jacobian at the origin are $-c/2 \pm \sqrt{c^2/4 - r}$, which are complex for $c^2 < 4r$. This permits the orbits to take on negative as well as positive values as they spiral in towards zero, which violates the requirement that $\phi \geq 0$. Therefore $c^2 \geq 4r$, and we have the situation portrayed in Figure 5.14 in which the dotted path indicates what an orbit would look like if it corresponded to a stationary wave front.

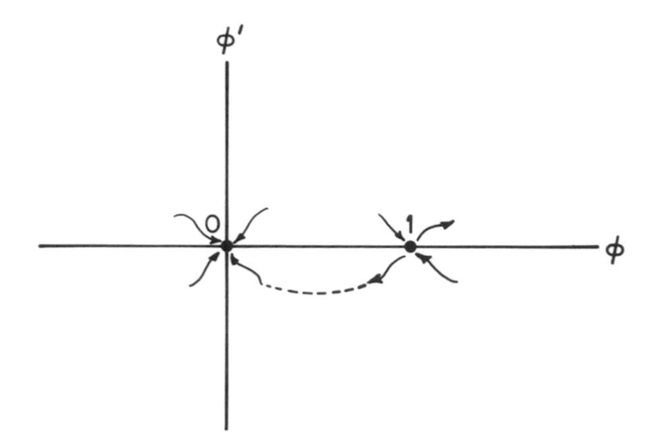

Figure 5.14 The ϕ, ϕ' phase plane corresponding to Equation 5.61 with $c \geq 2\sqrt{r}$. The equilibrium $\begin{pmatrix} 0 \\ 0 \end{pmatrix}$ is stable with a Jacobian whose eigenvalues are real and negative, while $\begin{pmatrix} 1 \\ 0 \end{pmatrix}$ is a saddle point.

Another type of wave solution occurring in some problems is a periodic wave train of period T. This has the form $\eta(x, t) = \phi(s)$ in which $s = kx - \omega t$. For any fixed t, the quantity k is the number of oscillations in an interval of length T. k is called the *wave number*, and T/k is the *wave length*. For x fixed, ω is the number of oscillations observed in a time period of duration T. ω is called the *frequency*. The

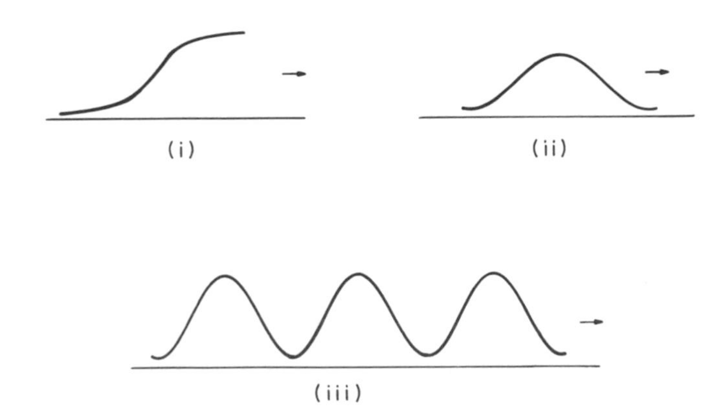

Figure 5.15 Various types of traveling wave solutions: (i) fixed wave front, (ii) solitary pulse, (iii) periodic pulse train.

wave train moves at a speed $c = \omega/k$ since $\phi(s) = \phi(k(x - ct))$ is a function of $x - ct$. A physical example occurs in the study of ocean waves (Section 5.7).

Finally there are traveling waves consisting of a solitary pulse (Exercise 5.8.14). Figure 5.15 illustrates the various possibilities.

5.6 Morphogenesis

The process whereby form and pattern evolve in a biological system is called morphogenesis. We will consider two simple models of morphogenesis in each of which there is initially an undifferentiated and spatially homogeneous distribution of some biological substance. This uniform distribution represents an unstable equilibrium state, and in both models it will be seen that a slight disturbance permits a nonhomogeneous spatial pattern to develop over time.

We are given a strip of tissue consisting of a homogeneous aggregate of living cells. Each cell has the same chemicals within it, and these diffuse through the tissue and react with one another. For simplicity, we consider only two such chemical agents. One is called a morphogen because in certain high enough concentrations it activates a change in cell structure and these altered cells are then induced to grow. The differentiated cells give rise to such recognizable features as hair follicles on skin or buds which sprout into shoots on plant stalks.

The other chemical constituent we call an inhibitor because it tends to dampen the growth of the morphogen. It is the interplay between the two chemicals and their movement that is responsible for the change in their concentrations. In Exercise 4.12, it was already seen how two chemicals can change over time as they react with one another. However, this process now occurs in space as well as in time, and we must stipulate certain conditions on the kind of interactions that are allowed to take place.

Let $\rho(x, t)$ be the morphogen concentration and $h(x, t)$ that of the inhibitor. The spatial variable denotes a one-dimensional slice of tissue consisting of cells, much as the slice of ocean in which plankton were located in Section 5.2. Although the cells are finite in size, we can assume that the chemicals concentrations are smooth functions by the same reasoning used in the traffic model earlier. In effect, the cells are shrunk to zero.

The interaction between the two chemical constituents either increases or diminishes. Their concentrations acts as source and sink terms in a diffusion model of type 5.11. These terms are functions of ρ and h, and therefore we can write down the equations

$$\frac{\partial \rho}{\partial t} = \mu \frac{\partial^2 \rho}{\partial x^2} + k_1(\rho, h) \tag{5.62}$$

$$\frac{\partial h}{\partial t} = \nu \frac{\partial^2 h}{\partial x^2} + k_2(\rho, h). \tag{5.63}$$

The diffusion coefficients μ and ν are deliberately chosen to be distinct. We do not specify the precise form of k_1 and k_2 because different variants exist in the literature. What they have in common, however, are three properties. First, the growth of the morphogen is *autocatalytic*, which is to say that it stimulates its own production. This means that $\partial k_1/\partial \rho > 0$. Second, the level of inhibitor rises with an increase in morphogen. We interpret this as $\partial k_2/\partial \rho > 0$. Finally the inhibitor suppresses the production of morphogens: $\partial k_1/\partial h < 0$. A set of reactions which satisfy these conditions are, for example, described by the relations

$$k_1(\rho, h) = \frac{a\rho^2}{h}$$

$$k_2(\rho, h) = c\rho^2$$

in which the constants are both positive.

Suppose that the chemicals have a spatially uniform distribution in which $\rho(x, t) = \bar{\rho}$ and $h(x, t) = \bar{h}$. These uniform densities satisfy 5.62 and 5.63 provided that $k_1(\bar{\rho}, \bar{h}) = k_2(\bar{\rho}, \bar{h}) = 0$. This defines an equilibrium solution to the equations. If ν/μ is large enough, then the equilibrium is unstable and gives rise to a spatially inhomogeneous pattern of ρ and h. This will be shown in some detail in the next example, but for the moment we content ourselves with an intuitive demonstration as to why this must be so. This is done in Figure 5.16. The inhomogeneous pattern contains peaks and troughs of morphogen concentration, and since the cells differentiate whenever the level of this chemical is large enough, it is clear that distinguishable features will appear among the aggregate of cells which make up the tissue wherever a peak occurs.

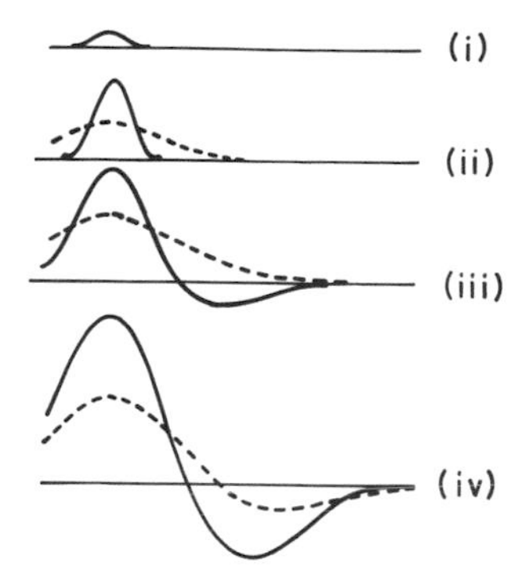

Figure 5.16 An intuitive demonstration of pattern formation. In (i) a uniform distribution (the straight line) is disturbed and there is a small increase in ρ. Because of autocatalysis ρ rises further in (ii). This is accompanied by an increase in h (dotted line). However, h spreads out faster since $\nu > \mu$. Then, in (iii), we see that wherever h is more abundant than ρ this forces ρ to decrease even further. A decrease in ρ eventually decreases h, as shown in (iv), and so ρ can begin to rise again. In this way, peaks and troughs eventually emerge.

Models leading to relations of the form 5.62 and 5.63 are said to be described by *reaction–diffusion* equations. Another example is Fisher's equation 5.59. A different morphogenic situation will now be discussed that leads to very similar equations. Recall the plankton moving in the sea within a favorable path of nutrients (Section 5.2). If the nutrient supply is uniformly distributed in the ocean, no noticeable bloom would be expected. However, a new element is introduced in terms of a predator, a tiny crustacean called a copepod, which grazes on the plankton. This is a predator–prey interaction of the type studied in Chapter Four except that now both species are allowed to move about in space. The plankton (or prey) of density ρ moves at random because of wave action while the copepods (the predators) of density h are self-propelling and disperse at will as they graze. The motion of each is considered to be diffusive. The dynamics of growth and decay of each species constitutes a set of source and sink terms for a diffusion model of the type considered above for chemicals in cells. In effect, the prey acts as a morphogen and the predator as an inhibitor and the question is whether a spatially homogeneous distribution of both predator and prey in the (one-dimensional slice of) ocean can eventually evolve into a heterogeneous pattern of prey density. Such a pattern would be seen as patches of plankton in the sea wherever ρ begins to crest. In this way, we have an alternate model of plankton patchiness without assuming favorable aggregates of nutrient. We ignore other factors that can affect growth such as sunlight or temperature or tidal effects. Each of these are assumed to be distributed evenly in the ocean.

Let us adopt the model equations 5.62 and 5.63 in which k_1 and k_2 are given by

$$\begin{aligned} k_1(\rho, h) &= r\rho\left(1 + \frac{\rho}{K}\right) - \alpha\rho h \\ k_2(\rho, h) &= -ch^2 + \beta\rho h. \end{aligned} \tag{5.64}$$

These relations are similar to the vector field of the quadratic model 4.21. There is an important new twist however. First, predator density decays quadratically in the absence of food due to severe competition. Second, prey density is autocatalytic: the per-capita growth rate of ρ is increased whereas it decreases in 4.21. This resembles the situation portrayed in Figure 4.2 where, for small densities at least, per-capita growth is also positive. It says that the plankton are self-generating, perhaps because of added protection from predators as their numbers increase or because the grazing efficiency of the predators declines with satiation.

To round out the model description assume that the ocean slice has a coast to coast width L. Since the marine organisms cannot migrate on shore, it is necessary to stipulate the boundary conditions

$$\frac{\partial \rho}{\partial x} = \frac{\partial h}{\partial x} = 0 \qquad \text{at } x = 0, L.$$

The complete reaction–diffusion model is then

$$\frac{\partial \rho}{\partial t} = \mu \frac{\partial^2 \rho}{\partial x^2} + r\rho\left(1 + \frac{\rho}{K}\right) - \alpha\rho h \tag{5.65}$$

$$\frac{\partial h}{\partial t} = \nu \frac{\partial^2 h}{\partial x^2} - ch^2 + \beta\rho h \tag{5.66}$$

in which the coefficients are positive. Since the copepods have greater mobility than the plankton the diffusion constant ν is taken to be greater than μ.

What we now show is that the different diffusive abilities of the species together with the autocatalytic growth of the prey suffices to destabilize an initially uniform distribution of plankton into a pattern of patchiness. The faster spread in space of the predator combined with the local self-activating growth of the prey does the trick. This is akin to pattern development of morphogens due to the wide dispersal of inhibitors reacting with the shorter range activation of morphogens.

A spatially uniform equilibrium of 5.65 and 5.66 that satisfies the boundary conditions occurs at $\bar{\rho}, \bar{h}$ where both k_1 and k_2 are zero in 5.64. This means that $\bar{\rho}, \bar{h}$ satisfy

$$\begin{aligned} \frac{r\rho}{K} - \alpha h + r &= 0 \\ \beta\rho - ch &= 0. \end{aligned} \tag{5.67}$$

$\bar{\rho}, \bar{h}$ is a unique positive equilibrium provided $\alpha\beta - rc/K > 0$ (Exercise 5.8.15). We assume this to be true from now on.

Let us linearize the nonlinear partial differential equations about the solutions $\rho = \bar{\rho}$, $h = \bar{h}$. By analogy with what we have been doing up to now for ordinary nonlinear differential equations, define perturbations u, v of the equilibrium by

$$\begin{aligned} \rho(x,t) &= \bar{\rho} + u(x,t) \\ h(x,t) &= \bar{h} + v(x,t). \end{aligned} \tag{5.68}$$

The nonlinear terms in 5.65 and 5.66 can be linearized about $\bar{\rho}, \bar{h}$ in the usual way, and so we obtain

$$\begin{aligned} \frac{\partial u}{\partial t} &= \mu \frac{\partial^2 u}{\partial x^2} + \frac{r\bar{\rho}u}{K} - \alpha\bar{\rho}v \\ \frac{\partial v}{\partial t} &= \nu \frac{\partial^2 v}{\partial x^2} + \beta\bar{h}u - c\bar{h}v. \end{aligned} \tag{5.69}$$

To be more specific, let us seek disturbances u, v that are proportional initially to $\cos \omega x$, for $\omega > 0$. The amplitude of the disturbance is made to depend exponentially on time. This gives

$$\begin{aligned} u(x,t) &= c_1 e^{\sigma t} \cos \omega x \\ v(x,t) &= c_2 e^{\sigma t} \cos \omega x, \end{aligned} \tag{5.70}$$

for suitable constants c_1 and c_2.

These perturbations are not as special as may first appear since a sufficiently mild but arbitrary perturbation may be Fourier represented as a sum of functions of type 5.70. Indeed, since we assume that $\partial\rho/\partial x$ and $\partial h/\partial x$ vanish at the endpoints of the interval, it is not hard to see that only the cosine terms are required in such a Fourier sum. If σ is small in magnitude, then the amplitude of the disturbance is also small. Substituting 5.70 into 5.69 we find that $e^{\sigma t}\cos\omega x$ cancels out leaving, in matrix form,

$$\sigma\begin{pmatrix} c_1 \\ c_2 \end{pmatrix} = \begin{pmatrix} \dfrac{r\bar{\rho}}{K} - \mu\omega^2 & -\alpha\bar{\rho} \\ \beta\bar{h} & -c\bar{h} - \nu\omega^2 \end{pmatrix}\begin{pmatrix} c_1 \\ c_2 \end{pmatrix}. \tag{5.71}$$

It follows that 5.70 satisfies the linearized system 5.69 provided that σ is an eigenvalue of the matrix

$$\Lambda = \begin{pmatrix} \dfrac{r\bar{\rho}}{K} - \mu\omega^2 & -\alpha\bar{\rho} \\ \beta\bar{h} & -c\bar{h} - \nu\omega^2 \end{pmatrix}. \tag{5.72}$$

To analyze this further, assume that the linearized system is stable to uniformly distributed disturbances, namely, those in which $\omega = 0$ in 5.70. This is assured if the matrix Λ in which $\omega = 0$ satisfies Det $\Lambda > 0$ and Trace $\Lambda < 0$ because it implies that $\sigma < 0$. Therefore, we require that

$$\frac{r\bar{\rho}}{K} - c\bar{h} < 0 \tag{5.73}$$

(Exercise 5.8.16). This too will be assumed from now on. It follows that if the uniform equilibrium is to be destabilized, the initial perturbation must be spatially uneven ($\omega \neq 0$). Returning to Λ when $\omega \neq 0$, it is apparent that Trace $\Lambda < 0$ because of condition 5.73. So for the stable distribution to become unstable to the disturbance, it must be that Det $\Lambda < 0$. Now

$$\text{Det}\,\Lambda = \left(\alpha\beta - \frac{rc}{K}\right)\bar{\rho}\bar{h} + \left(\mu c\bar{h} - \frac{\nu r\bar{\rho}}{K}\right)\omega^2 + \mu\nu\omega^4. \tag{5.74}$$

For a unique positive solution to 5.69, we assumed $\alpha\beta - rc/K > 0$. Therefore, Det $\Lambda > 0$ when $\omega = 0$. Since the ω^4 term dominates when ω

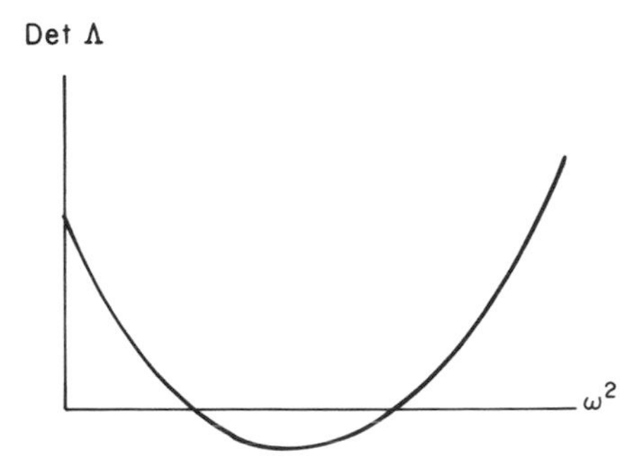

Figure 5.17 Graph of the determinant of Λ in 5.74 as a function of ω^2.

is large, it also follows that Det Λ is positive for large enough ω. In order for the determinant to ever be negative at some intermediate ω value, the quadratic function in ω^2 given by 5.74 must have two positive roots (Figure 5.17). This requires that

$$\frac{r\nu\bar{\rho}}{K} - \mu c\bar{h} > 0$$

and

$$\left(\frac{r\nu\bar{\rho}}{K} - \mu c\bar{h}\right)^2 > 4\left(\alpha\beta - \frac{rc}{K}\right)\bar{\rho}\bar{h}\mu\nu.$$

Using the first of these inequalities to take a positive square root in the second leads to

$$\sqrt{\frac{\nu}{\mu}}\left(\frac{r\bar{\rho}}{K}\right) - \sqrt{\frac{\mu}{\nu}}\,(c\bar{h}) > 2\sqrt{\left(\alpha\beta - \frac{rc}{K}\right)\bar{\rho}\bar{h}}\,. \tag{5.75}$$

We know that $\bar{\rho}$ and $\bar{h}$ are positive if $\alpha\beta - rc/K > 0$, and so the expression under the radical in 5.75 must be positive. Therefore, the inequality is satisfied provided ν/μ is large enough. Since it has been assumed that ν is larger than μ, this is a plausible requirement. It follows that a spatially heterogeneous pattern is initiated for some intermediate range of ω values. Observe that the boundary conditions $\partial\rho/\partial x = \partial h/\partial x = 0$ require $\sin \omega x$ to be zero at $x = 0, L$. Therefore, L must satisfy $\omega L = k\pi$ for $k = 1, 2, \ldots$. Let $r_1 < r_2$ be the two positive roots of Det Λ as a quadratic in ω^2. For Det $\Lambda < 0$, one must then have

$r_1 < (k\pi/L)^2 < r_2$ or, to put it another way, $\pi k/\sqrt{r_2} < L < \pi k/\sqrt{r_1}$. Hence, a solution to the boundary value problem exists only if L is large enough. Moreover, given L, only finitely many k will satisfy the above inequality, and so the system is excited, rather than damped, only for those perturbations having specific spatial frequencies.

Unfortunately, we have carried out a linearized stability analysis, and so it is not possible to conclude immediately how the perturbation will grow. However, some understanding of what ultimately happens can be obtained by trying to find a steady state solution of 5.65 and 5.66 in the limiting case of $\mu = 0$. This limiting situation approximates the situation when ν/μ is very large. The steady state, or time independent, equations become

$$r\rho\left(1 + \frac{\rho}{K}\right) - \alpha\rho h = 0$$
$$\nu\frac{d^2h}{dx^2} - ch^2 + \beta\rho h = 0. \tag{5.76}$$

If ρ is not to be identically zero, the first of these relations gives

$$\rho = \frac{K(\alpha h - r)}{r}.$$

Substituting into the second equation of 5.76 results in a second-order ordinary differential equation of the type $p'' + f(p) = 0$, which was studied extensively by phase plane methods in Chapter Two:

$$\frac{d^2h}{dx^2} + \frac{h^2}{\nu}\left(\frac{\alpha\beta K}{r} - c\right) - \frac{K\beta h}{\nu} = 0.$$

To simplify matters, relabel the coefficients of h^2 and h to be positive constants a and b (a is positive by an earlier assumption). Then

$$\frac{d^2h}{dx^2} + ah^2 - bh = 0. \tag{5.77}$$

As in Section 5.2, it is useful to carry out a phase plane analysis by considering the conservation of "energy" equation

$$\frac{(h')^2}{2} + U(h) = C,$$

where C is a constant and the potential is given as

$$U(h) = a\int_0^h s\,ds - bs\int_0^h ds = \frac{ah^3}{3} - \frac{bh^2}{2}. \tag{5.78}$$

This has the same form as 5.25 except that it is of opposite sign. Written as a first order system, 5.77 reveals two equilibria at $\begin{pmatrix} h \\ h' \end{pmatrix} = \begin{pmatrix} 0 \\ 0 \end{pmatrix}$ and $\begin{pmatrix} b/a \\ 0 \end{pmatrix}$. The first is an unstable saddle, and the other neutrally stable.

Since the reader has carried out the phase plane analysis in Section 5.2, it suffices to refer to Figure 5.18 to interpret what happens. We are dealing with a coast to coast slice of ocean of length L. Since our sea creatures do not migrate on shore, we know that

$$h'(0) = h'(L) = 0.$$

This means that the orbits in Figure 5.18 begin and end on the horizon-

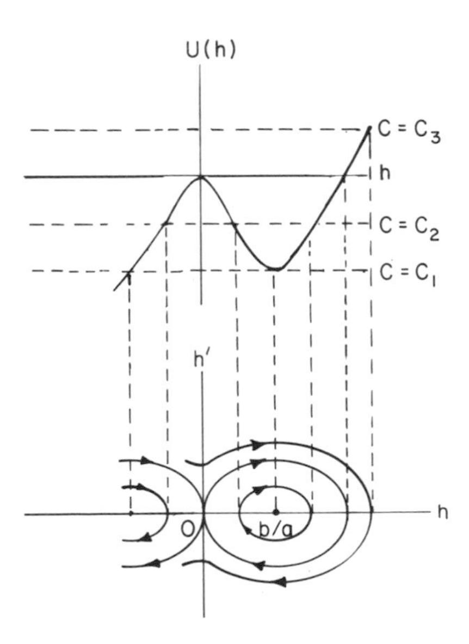

Figure 5.18 Phase portrait of Equation 5.77. Orbits are shown for different "energy levels" C. The points where h equals zero and b/a on the horizontal axis are equilibrium states. This phase diagram is a mirror image of that in Figure 5.1.

tal h axis. If L is infinite, this corresponds to an orbit along the stable manifold that leads to the saddle point. Otherwise, the orbits must lie inside the separatrix and consist of integer multiples of half cycles in which ρ peaks a finite number of times in a periodic way. This would result in a regularly repeating pattern of plankton patches, a questionable conclusion which is perhaps unavoidable given all the assumptions made. But it does at least show that a heterogeneous pattern is possible. In terms of the cellular model, a periodically repeating spatial distribution of activated cells is more understandable since this is often observed in nature (see Exercise 5.8.17).

The energy conservation relation shows, as before, that

$$h'(x) = \pm\sqrt{2[C - U(h)]}. \tag{5.79}$$

The orbits inside the separatrix are symmetrical about the ρ axis, and so they cover the same distance l in each half cycle. Assuming that $h' \neq 0$ (for otherwise $h(x)$ is a constant, the uniform steady state), we obtain

$$l = \int_0^l dx = \int_{h_1}^{h_2} \frac{dh}{h'} = \int_{h_1}^{h_2} \frac{dh}{\sqrt{2[C - U(h)]}}, \tag{5.80}$$

where $0 < h_1 < h_2 < b/a$ are the two points where the orbit crosses the horizontal axis. The energy relation tells us that $U(h) < C$ with $U(h_1) = U(h_2) = C$ (since the derivative of h is zero on the ρ axis). Moreover, $h'(x) \geq 0$ as h moves from h_1 to h_2. Therefore, the quantity inside the square root in 5.80 is nonnegative, and we can take the positive sign of this root. On the other hand, if the orbit moves on the half cycle from ρ_2 to ρ_1, then $\rho' \leq 0$, but the interchange of integration limits again results in the same integral. Therefore,

$$l = \int_{h_1}^{h_2} \frac{dh}{\sqrt{2[U(h_2) - U(h)]}}.$$

As in the algae model of Section 5.2, it is necessary to caution that a solution to 5.77 does not exist if L is too small. In fact by linearizing 5.77 about $\begin{pmatrix} b/a \\ 0 \end{pmatrix}$, we see that in a small neighborhood of this point the equation is roughly that of the harmonic oscillator $h'' + bh = 0$. Its

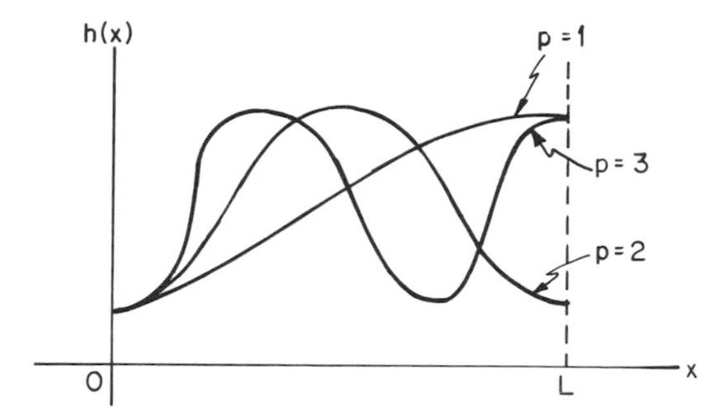

Figure 5.19 Successive plots of h for the system 5.77 as an orbit completes one, two, or three cycles on an interval of length L. Note that $h'(0) = h'(L) = 0$.

period of oscillation is $2\pi/\sqrt{b}$, and so L cannot be less than a single half period $\pi/\sqrt{b}$. This is the same conclusion reached earlier in considering the range of values for which Det $\Lambda < 0$. In fact by letting $\mu = 0$ in the expression for Det $\Lambda = 0$, one obtains a single root $r = \omega^2$ whose value equals b since $\bar{h} = a/b$.

In general, if the orbit completes p half cycles, the total length L must equal pl. Let us therefore suppose that a spatially non-uniform steady state density h exists on an interval of length L. By virtue of the linear relationship between ρ and h a similar density also exists for ρ. From Figure 5.18, it is apparent that as h moves from h_1 to h_2 then h' at first increases and then decreases. The opposite occurs as h moves in the reverse direction. This results in a wave of evenly spaced peaks and troughs as p half cycles are completed. Since $h' = 0$ at the endpoints, the wave reaches a maximum or minimum at 0 or L. The same is, of course, true for ρ (Figure 5.19).

5.7 Tidal Dynamics

The water quality model of Section 5.3 treated a river flowing steadily with a constant average velocity c along a longitudinal direction. c is generally easy to estimate. It would be useful to extend this analysis to other situations in which the flow velocity u is not constant. What we have in mind are other water bodies in which the flow is in a roughly one-dimensional direction, such as canals, narrow bays, and estuaries that are connected to the ocean at one end and therefore subject to tidal oscillations. In these cases, velocity can vary in magnitude as well as in direction.

The concentration of pollutants (or organic nutrients or even dissolved oxygen) that are carried in the water body was characterized earlier by a simple advective equation. In the presence of tides, however, there would also be diffusive effects due to the back and forth movement of the water. This folds the pollution plume onto itself thereby dispersing it. Therefore an advection–diffusion equation is more appropriate. The dispersion coefficient ν may itself vary with x and t, and a straightforward modification of the argument that led to Equation 5.10 now gives

$$\frac{\partial \rho}{\partial t} = -u\frac{\partial \rho}{\partial x} + \frac{\partial}{\partial x}\left(\nu\frac{\partial \rho}{\partial x}\right) + k \tag{5.81}$$

for the pollution concentration ρ. ν is assumed to be known empirically, and often it can be approximated by a constant, but u is unknown and must be determined. In general, u is independent of ρ. Moreover, it can vary in sign, depending on tidal direction. The determination of u is accomplished by finding a pair of differential equations that u must satisfy. Why, one may ask, did we wait until now to do so and not introduce this material earlier at the end of Section 5.3? The answer is that in the present situation we are obliged to use a different modeling principle than that of mass conservation.

The density of water (mass per unit volume) is taken to be a constant, a value that is denoted as $\hat{\rho}$ to distinguish it from the use of ρ as mass per unit length. The water body is approximated by a channel of constant width b, and we consider what happens to a cross-sectional slice of length Δx in the longitudinal direction. The mass Δm of water within the slice is then $\hat{\rho}bh\,\Delta x + o(\Delta x)$, where h is water elevation above the horizontal channel bottom. It follows that ρ, the mass per unit length, is $\hat{\rho}bh$. The volume per unit time flowing across the face of the cross-sectional slice is bhu, where u is the longitudinal velocity. Both u and h are nonconstant functions of x and t.

Observe that while earlier we considered the change in concentration of a pollutant (or, in Section 5.3, of algae) within a water body, our interest has now shifted to changes in the body of water itself.

With this information one can apply the principle of mass conservation (Equation 5.3) to obtain the following specific relation between the

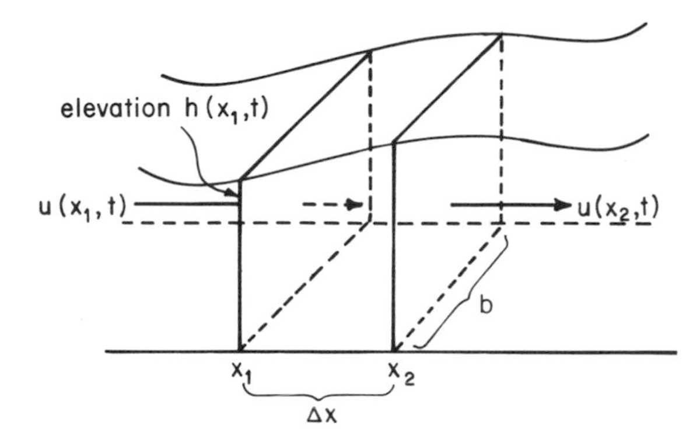

Figure 5.20 A slice of water of varying elevation h, constant width b, and thickness Δx.

variables u and h:

$$\frac{\partial h}{\partial t} + h\frac{\partial u}{\partial x} = -u\frac{\partial h}{\partial x}. \tag{5.82}$$

The derivation of 5.82 from 5.3 is left to the reader (Exercise 5.8.18).

Since 5.82 involves two unknowns, we need a second relation in order to solve for either the elevation h or velocity u. This will be done by invoking Newton's law of conservation of momentum, which states that the time rate of change of momentum of a particle must equal the sum of forces acting on it. This principle is applied to the cross-sectional slice of water of thickness Δx between x_1 and x_2 as shown in Figure 5.20. The entire slice is treated as a single particle.

Since momentum is the product of mass and velocity, the momentum of the slice is given by

$$\hat{\rho} b \int_{x_1}^{x_2} h(x,t)\, u(x,t)\, dx. \tag{5.83}$$

By Taylor's Theorem,

$$\begin{aligned} u(x_2,t) &= u(x_1,t) + \frac{\partial u(x_1,t)}{\partial x}\Delta x + o(\Delta x) \\ h(x_2,t) &= h(x_1,t) + \frac{\partial h(x_1,t)}{\partial x}\Delta x + o(\Delta x). \end{aligned} \tag{5.84}$$

The end points x_1 and x_2 move as time evolves. By definition,

$$\begin{aligned} \dot{x}_1 &= u(x_1, t) \\ \dot{x}_2 &= u(x_2, t). \end{aligned} \tag{5.85}$$

The time rate of change of 5.83 can now be computed by utilizing the rule 5.53 for differentiation of integrals with variable limits. Applying 5.53 to 5.83 then gives, by virtue of 5.84 and 5.85,

$$\begin{aligned} &\hat{\rho} b \left\{ \int_{x_1}^{x_2} \frac{\partial (hu)}{\partial t} dx + \dot{x}_2 h(x_2, t)\, u(x_2, t) - \dot{x}_1 h(x_1, t)\, u(x_1, t) \right\} \\ &\quad = \hat{\rho} b \left\{ \int_{x_1}^{x_2} \left(h \frac{\partial u}{\partial t} + u \frac{\partial h}{\partial t} \right) dx \right. \\ &\qquad \left. + \left[uh \frac{\partial u}{\partial x} + u \left(h \frac{\partial u}{\partial x} + u \frac{\partial h}{\partial x} \right) \right] \Delta x \right\} + o(\Delta x), \end{aligned} \tag{5.86}$$

where all the terms on the right are evaluated at (x_1, t). But from 5.82, we see that

$$\frac{\partial h}{\partial t} = -\left(h \frac{\partial u}{\partial x} + u \frac{\partial h}{\partial x} \right). \tag{5.87}$$

Substituting 5.87 into 5.86 gives

$$\begin{aligned} &\hat{\rho} b\, \Delta x \left[\left(h \frac{\partial u}{\partial t} + u \frac{\partial h}{\partial t} \right) + uh \frac{\partial u}{\partial x} - u \frac{\partial h}{\partial t} \right] + o(\Delta x) \\ &\quad = \hat{\rho} b h\, \Delta x \left(\frac{\partial u}{\partial t} + u \frac{\partial u}{\partial x} \right) + o(\Delta x) \end{aligned} \tag{5.88}$$

The expression 5.88 for the time rate of change of momentum must equal the sum of forces acting on the slice. One of these, which we ignore, is an empirical relation that stipulates a frictional force resisting fluid flow due to the roughness of the channel walls. In shallow wide canals or in deep narrow ones, the frictional force can be noticeable and would have to be included.

Another force is that of water pressure. The pressure force of water at a depth $h - z$ measured from the top is known to be $\bar{\rho} g(h - z)$ and so the total pressure force on the face of cross-sectional width b at x is

$$P(x_1, t) = \bar{\rho} b g \int_0^{h(x,t)} [h(x,t) - z]\, dz.$$

Since

$$P(x_2, t) = P(x_1, t) + \frac{\partial P(x_1, t)}{\partial x} \Delta x + o(\Delta x),$$

the difference in pressure between the two faces of the slice is

$$P(x_1, t) - P(x_2, t) = -\frac{\partial P(x_1, t)}{\partial x} \Delta x + o(\Delta x).$$

Utilizing 5.53 again we see that the net pressure on the water slice is, up to terms in $o(\Delta x)$,

$$\bar{\rho} b g\, \Delta x \left\{ -\frac{\partial}{\partial x} \int_0^{h(x,t)} [h(x,t) - z]\, dz \right\} \Bigg|_{x = x_1}$$

$$= -\bar{\rho} b g\, \Delta x \int_0^{h(x_1,t)} \frac{\partial h}{\partial x}\, dz = -\bar{\rho} b g\, \Delta x\, h \frac{\partial h}{\partial x}. \tag{5.89}$$

Combining the pressure force with 5.88 and letting Δx tend to zero gives, finally, the conservation of momentum equation

$$\frac{\partial u}{\partial t} + u \frac{\partial u}{\partial x} = -g \frac{\partial h}{\partial x}. \tag{5.90}$$

The nonlinear equations 5.82 and 5.90 can be solved together to yield h and u for suitable initial and boundary conditions.

However, it is not possible to obtain an explicit solution in general, and one must resort to numerical techniques or the use of approximations. In order to illustrate the second approach, we consider a very special case. Let $\bar{h}$ be the average water depth. Then $h = \bar{h} + \eta$, in which the variation in wave amplitude η is assumed to be small compared to h. Also suppose that the waves are long so that the change in η and u is gradual. We interpret this to mean that u and the partial derivatives of η with respect to x and t are not too large and that products of these terms can therefore be neglected. Equation 5.82 can then be approximated by

$$\frac{\partial \eta}{\partial t} = -\bar{h}\frac{\partial u}{\partial x}. \tag{5.91}$$

From this we see that $\partial u/\partial x$ is not too large, and so $u\,\partial u/\partial x$ can also be neglected. The momentum equation 5.90 therefore reduces to

$$\frac{\partial u}{\partial t} + g\frac{\partial \eta}{\partial x} = 0. \tag{5.92}$$

Differentiating 5.91 with respect to t and 5.92 with respect to x gives

$$\frac{\partial^2 \eta}{\partial t^2} = c^2\frac{\partial^2 \eta}{\partial x^2} \tag{5.93}$$

in which $c = \sqrt{g\bar{h}}$. This is known as the *wave equation.*

It is known that the most general solution to Equation 5.93 can be written as the sum of two traveling waves, one moving to the right, the other to the left (Exercise 5.8.19). It is therefore not unreasonable that a periodic wave train of the form

$$\eta(x, t) = a\cos(kx \pm \omega t)$$

also satisfies the wave equation. In fact $\eta(x, t) = a\cos(k(x \pm ct))$, and so it is a function of $x \pm ct$, where $c = \omega/k$. That this satisfies 5.93 is easily verified by substitution. The quantity $2\pi/k$ is the wave length, and ω is the frequency of oscillation. We see that η is a traveling wave whose velocity is c. Thus, an observer riding with the wave at velocity c always sees the same wave height.

Now consider the situation in which the channel is closed at one end (at location $x = 0$) and is open to the ocean at $x = L$. This represents the case of a long bay or estuary of fairly uniform width (like Long Island Sound, near New York City) that is subject to the movement of the ocean tides at only one end. Incidently, the period $2\pi/\omega$ of a tidal wave is nearly twelve hours, and at a mean evaluation $\bar{h}$ of fifty feet, the relation $c = \sigma/k = \sqrt{g\bar{h}}$ can be used to find a wavelength $2\pi/k$ of about 350 miles. This is much longer than the length of a typical channel.

In any event, suppose that η_m is the maximum wave amplitude as measured, for example, at the closed end.

The periodic long tidal wave enters the channel at the ocean end, and we assume that it is completely reflected at the closed end. Hence, the water elevation can be represented as a sum of two periodic traveling waves of the same wavelength and frequency, each moving in the opposite direction:

$$\eta(x, t) = \alpha_1 \cos(kx - \omega t) + \alpha_2 \cos(kx + \omega t). \qquad (5.94)$$

Choose high tide at $x = 0$ to occur at time $t = 0$. This is given as η_{m}. Hence $\alpha_1 + \alpha_2 = \eta_{\mathrm{m}}$. Since the incident and reflected waves at $x = 0$ must be the same for all t, it follows that

$$\alpha_1 \cos(-\omega t) = \alpha_2 \cos \omega t.$$

That is, $\alpha_1 = \alpha_2$. Now $\cos(kx \pm \omega t) = \cos kx \cos \omega t \mp \sin kx \sin \omega t$, and so relation 5.94 reduces to

$$\eta(x, t) = \eta_{\mathrm{m}} \cos kx \cos \omega t. \qquad (5.95)$$

From 5.91, we obtain

$$\frac{\partial u}{\partial x} = -\frac{1}{\bar{h}} \frac{\partial \eta}{\partial t} = \left(\frac{\eta_{\mathrm{m}} \omega}{\bar{h}} \right) \cos kx \sin \omega t.$$

Integrating this expression, it follows that

$$u(x, t) = \left(\frac{\eta_{\mathrm{m}} c}{\bar{h}} \right) \sin kx \sin \omega t, \qquad (5.96)$$

and so amplitude and velocity are out of phase. Let l denote one quarter of a wavelength. Then $\eta(\lambda l, t) = 0$ for all t when $\lambda = 1, 2, 3, \ldots$. Hence, the solution 5.95 is actually a standing wave that moves only up and down. If $L < l$, high or low tide occurs simultaneously at all points.

The problem gets more complicated when separate tidal conditions are imposed at each boundary. An example would be a canal that connects two oceans (such as a proposed sea-level Panama Canal). We leave the formulation and solution of this problem as an interesting modeling exercise (Exercise 5.8.20).

5.8 Exercises

5.8.1 In deriving Equation 5.10, why is it that the density ρ of a substance and not its mass appears?

5.8.2 Do a linearized stability analysis of the equilibrium states of system 5.26.

5.8.3 A country wishes to establish a no-fishing zone along its coastal waters to protect some species of fish which grows exponentially. Outside the zone of width L, all the fish are harvested by deep sea trawlers. The fish move about at random in the direction from shore to the zone's boundary. If L is too small, the destruction of the fish may exceed their ability to reproduce new recruits, which would lead to eventual extinction. Find the minimum zone width L that prevents this from happening by formulating a suitable model. The fish cannot swim on-shore so be careful about boundary conditions.

5.8.4 Show that the integral 5.28 has the value $\pi\sqrt{\nu/r}$ when $U(\rho) = \rho^2/2$.

5.8.5 The growth of a population of insects at time t depends on the age distribution of its members (that is, fertility is a function of age). Let age be denoted by $a \geq 0$, and let $\rho(a, t)$ be the population density at age a and time t. Find a partial differential equation satisfied by ρ. Suppose that the population growth rate is determined by

$$\frac{d\rho(a, t)}{dt} = r\rho(a, t)$$

for each $a \geq 0$ (r is constant). Interpret the meaning of

$$\rho(0, t) = r\int_0^\infty \rho(a, t)\, da.$$

5.8.6 Show that a traveling wave solution to 5.30, subject to boundary condition 5.34 is

$$v(x, t) = \gamma U\left(t - \frac{x}{c}\right).$$

5.8.7 Find the steady state solution $\delta(x)$ to Equation 5.37 using boundary conditions 5.34 and 5.38. Sketch the solution and interpret its meaning. This is known as the *Streeter–Phelps oxygen-sag curve*.

5.8.8 Let $\rho = \bar{\rho}$ be an average car density on a highway. Suppose $\rho = \bar{\rho} + v$ where v is small enough so that a linearization of $q(\rho)$ about $\bar{\rho}$ is reasonable. Show that v satisfies the linear equation 5.30.

5.8.9 Solve Equation 5.50 with the initial condition $x = x_0$ at $t = -x_0/u_m$, $x_0 < 0$. Determine how long it takes for the car that starts at x_0 to reach the light at $x = 0$. Compare this to a situation in which the car at x_0 has no traffic in front of it. What happens as $t \to +\infty$?

5.8.10 Suppose initial traffic density is

$$\rho(x, 0) = \begin{cases} \rho_1 < \rho_m, & x > 0, \\ 0, & x < 0. \end{cases}$$

Determine the shock velocity, draw the characteristics, and interpret the result physically.

5.8.11 Show that the shock velocity $\dot{x}_s$ in 5.55 is the same as the local wave velocity $dq/d\rho$ evaluated at the average density $(\rho_+ + \rho_-)/2$. If one rescales by letting $\psi = (1 - 2\rho/\rho_m)u_m$, then traffic equation becomes

$$\frac{\partial \psi}{\partial t} = \psi \frac{\partial \psi}{\partial x}$$

and $\dot{x}_s = (\psi_+ + \psi_-)/2$.

5.8.12 Derive Burger's equation 5.58 from 5.57.

5.8.13 Show that Equation 5.58 has a traveling wave solution of the form $\rho(s)$ with $s = x - ct$, $c > 0$, in which ρ is monotone increasing from zero to some $\rho_1 \le \rho_m$. To simplify the analysis, it is convenient to rescale Burger's equation by letting $\psi = u_m(1 - 2\rho/\rho_m)$. Then 5.58 becomes

$$\frac{\partial \psi}{\partial t} + \psi \frac{\partial \psi}{\partial x} = \nu \frac{\partial^2 \psi}{\partial x^2}. \tag{5.97}$$

Interpret the result physically as $\nu \to 0$, and compare this limiting case to the specific situation posed in Exercise 5.8.10. Show, in fact, that the wave speed c is identical to the shock velocity $\dot{x}$ as $\nu \to 0$. Why is this reasonable?

5.8.14 In Burger's equation 5.97, the nonlinear term $\psi\, \partial\psi/\partial x$, which by itself induces a shock wave, is balanced by the diffusion term. This permits a wave front to form that does not break (see Exercise 5.8.13). In some problems, the nonlinear effect of shock can be compensated for in another way. To see this, consider a linear equation

$$\frac{\partial \psi}{\partial t} + \nu \frac{\partial^3 \psi}{\partial x^3} = 0.$$

If we attempt a traveling wave solution of the form

$$\phi(x, t) = \phi(x - ct) = \cos\left(k\left(x - \frac{\omega t}{k}\right)\right), \qquad c = \frac{\omega}{k},$$

it is quickly found that this satisfies the equation provided $\omega = \nu k^3$. As k varies so does ω, and therefore different wave lengths propagate at different speeds c (see the discussion in Section 5.5). Since the equation is linear, the most general solution is a superposition of such terms, one for each k, which means in effect that the solution consists of waves traveling at different speeds. That is, they disperse.

Consider now the following quasilinear equation, known as the *Korteweg–deVries equation*:

$$\frac{\partial \psi}{\partial t} = \psi \frac{\partial \psi}{\partial x} + \nu \frac{\partial^3 \psi}{\partial x^3}. \tag{5.98}$$

The nonlinear term which tends to produce a shock front is now compensated by the dispersion term $\nu \dfrac{\partial^3 \psi}{\partial x^3}$. It plays a role similar to that of diffusion in 5.97. To see this attempt a traveling wave solution of the form $\psi(x, t) = \phi(s)$ with $s = x - ct$. Show that

$$\phi(s) = a + b\,\mathrm{sech}^2 \sqrt{\frac{bs}{12k}}$$

is a solution to 5.98 for positive constants a and b and wave speed $c = a + b/3$. Describe the properties of the wave front ϕ, which is known as a *soliton*.

5.8.15 Establish that Equations 5.67 have a unique positive solution if $\alpha\beta > rc/K$.

5.8.16 Show that condition 5.73 suffices to insure that eigenvalues of matrix Λ in 5.72 are negative (when $\omega = 0$).

5.8.17 Carry out a steady state analysis of the cellular development model 5.62 and 5.63 in the case in which

$$k_1(\rho, h) = a(\rho - \bar{\rho}) - b(h - \bar{h})$$

$$k_2(\rho, h) = c(\rho - \bar{\rho})$$

with a, b, c positive. The homogeneous solution therefore occurs at $\bar{\rho}$, $\bar{h}$. Assume ν is large relative to μ by making $\mu = 0$. Consider a ring of cells circling about the stem of a vine, for example, of circumference L. The value of ρ and h at zero and L must therefore be the same. Show that a nonhomogeneous steady state exists as a periodic solution $\rho(x)$. This offers a possible explanation for the whorling of leaves about a stem, in which each leaf is rotated a fixed angle clockwise from its neighbor at each point where the concentration ρ peaks.

Do this in two ways. First, by an argument similar to that used in Section 5.6 and, secondly, by explicitly obtaining the steady state solution. This morphogenesis model is due to A. Turing, who is perhaps better known for his work on computability and "Turing machines."

5.8.18 Derive Equation 5.82 of conservation of mass in tidal flow from Equation 5.3.

5.8.19 Show that the general solution to the wave equation 5.93 is a sum of two traveling waves along the x axis, one moving to the right, the other to the left, each at speed c. That is,

$$\phi(x, t) = f(x - ct) + g(x + ct)$$

for suitable functions f and g.

5.8.20 A canal connects two oceans at $x = 0$ and $x = L$. Each ocean has a different periodic tidal oscillation at the canal boundaries. Formulate and solve this boundary value problem for the tidal elevation η.

Hint: The tides at one end do not disturb the water level in the other ocean and only affect the water elevation within the canal. In effect each tide is completely reflected off the opposite end of the canal.

Chapter Six

Cycles and Bifurcation

6.1 Self-Sustained Oscillations

We have seen that if the equilibrium of a linear dynamical system in $\mathbf{R}^2$ is unstable, then any orbit not on its stable manifold moves increasingly toward infinity. Instability is essentially a global affair. For nonlinear systems, the situation is less obvious. An unstable equilibrium forces most orbits to move away in the vicinity of the rest point. If an orbit is repelled by the unstable point, what happens to it? The answer is that it could tend to a new and stable configuration. As will be seen in this chapter, it is indeed possible for an orbit to approach a closed cycle that surrounds the unstable point: local instability may give rise to a globally stable cycle. A version of this phenomenon was provided by the study of reaction–diffusion equations in the previous chapter in which a spatially uniform and unstable equilibrium generates a spatially non-uniform but stable pattern.

In order to explain the limiting behavior about unstable points, we proceed by discussing several examples.

Example 6.1 Dry or coulomb friction in a mass–spring system was discussed in Section 1.1. Let us now investigate a more interesting

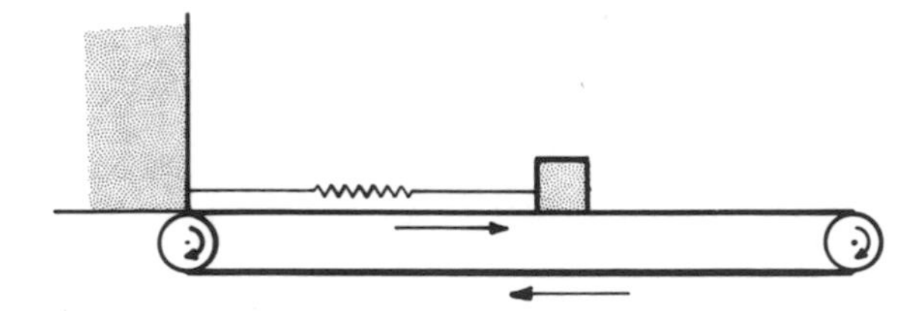

Figure 6.1 A mass–spring system atop a frictional belt moving to the right with speed v.

situation in which a block of mass m is attached to a spring and placed on a sticky surface that itself moves to the right at a constant speed v. The block is on a conveyer belt to which it adheres for awhile until the restoring force of the spring pulls it back (Figure 6.1).

Two situations come to mind in which this model might be applicable. In one, a resinated bow is stroked transversely across a violin string, and the other is a bicycle wheel rim being pressed against a brake pad when the hand grip is squeezed. In each case a sound emerges (which in some cases may be a squeal rather than a tone). Sounds come about from vibrations in the surrounding air, and so we expect that some sort of back and forth motion must have been set up on the violin string or brake pad. Each of these is flexible enough to be displaced a little as it adheres to the bow or wheel rim, but being tautly held to a rigid support there is a force tending to restore their positions to rest, much as in the mass–spring configuration. The violin string, incidently, is treated here as a point mass which effectively means that its behavior is uniform along the entire string length.

A mathematical description of the motion was provided in Section 1.1 except that there the surface is not moving. With a moving belt, it is convenient to consider the velocity of the block relative to v. As long as the block is held gripped by the sticky surface relative motion is zero: $\dot{x} = v$. In this event, the restoring force exerted by the spring is exactly matched by the frictional force of the belt. This means that for some $\eta > 0$ the frictional force is kx when $|kx| \leq \eta$. However, once motion ensues ($\dot{x} \neq v$) the frictional force is given by

$$g(\dot{x} - v) = -k_1 \operatorname{sgn}(\dot{x} - v). \tag{6.1}$$

The constant k_1 is less than η since it takes less to keep the body in motion than it does to dislodge it in the first place.

Let $k/m = \omega^2$ and $k_1/m = \alpha$. Then, for $\dot{x} - v \neq 0$, Equation 1.4 gives

$$\ddot{x} + \omega^2 x + \alpha \operatorname{sgn}(\dot{x} - v) = 0. \tag{6.2}$$

A phase plane analysis of this equation was carried out in Exercise 2.5.8 for the non-moving belt. A similar argument applies in the present situation. In essence, there are two equations in 6.2 depending on the sign of $\dot{x} - v$. The total energy associated with each is of course constant:

$$\frac{\dot{x}^2}{2} + \frac{\omega^2 x^2}{2} \pm \alpha x = \text{constant},$$

which can be rewritten, with a different constant, as

$$\dot{x}^2 + \omega^2 \left(x \pm \frac{\alpha}{\omega^2} \right)^2 = \text{constant}. \tag{6.3}$$

Relation 6.3 describes two families of ellipses in the x, $\dot{x}$ phase plane, one centered at $\begin{pmatrix} \alpha/\omega^2 \\ 0 \end{pmatrix}$ and the other at $\begin{pmatrix} -\alpha/\omega^2 \\ 0 \end{pmatrix}$. In addition, the orbit moves to the right on the horizontal $\dot{x} = v$ axis for $|x| \leq \eta/k$. Observe that $\alpha/\omega^2 = k_1/k < \eta/k$. The direction of the arrows in the phase plane are determined by the fact that $\dot{x} > 0$ as x increases. We note, parenthetically, that Figure 6.2 can be redrawn, if one wishes, as families of circles in an x, $\dot{x}/\omega$ phase plane by writing 6.3 as

$$\left(\frac{\dot{x}}{\omega} \right)^2 + \left(x \pm \frac{\alpha}{\omega^2} \right)^2 = \text{constant}.$$

Figure 6.2 shows that the orbits above and below the line $\dot{x} = v$ join up, for $|x| \leq \eta/\omega$, with the orbit moving left to right. The tangent to these orbits jumps at each point along this line.

Observe the family of cycles that exist about $\begin{pmatrix} \alpha/\omega^2 \\ 0 \end{pmatrix}$. Especially interesting is the orbit which begins at $\begin{pmatrix} \eta/k \\ v \end{pmatrix}$. Call it Γ. Suppose the block is initially placed on the belt between $\pm\eta/k$. It adheres to the sticky surface and moves to the right at velocity v. Beyond η/k the restoring force of the spring exceeds that of friction, and so the block

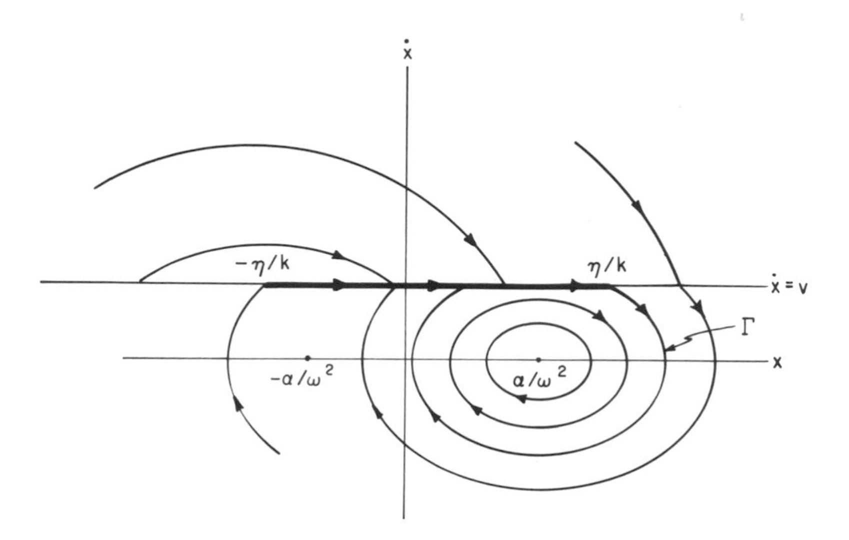

Figure 6.2 Phase portrait of Equation 6.2 for a mass–spring system on a moving belt. There are two families of elliptical orbits depending on the sign of $\dot{x} - v$, one centered at $\begin{pmatrix} \alpha/\omega^2 \\ 0 \end{pmatrix}$ in the $x, \dot{x}$ plane, and the other at $\begin{pmatrix} -\alpha/\omega^2 \\ 0 \end{pmatrix}$.

begins to slip with a velocity $\dot{x} < 0$. The net speed is $\dot{x} + v$ and so it continues to the right beyond η/k until $\dot{x} = -v$ at which point it is seen to move backward reaching a maximum velocity at α/ω^2. Thereafter it slows down due to a decreasing restoring force until $\dot{x} = -v$ again. It then starts to move to the right until $\dot{x} = v$, at which juncture the moving belt once more grabs the block. A cycle has been initiated that repeats itself indefinitely along Γ. What has happened is that the energy dissipation due to friction is counteracted by the external energy supplied by a moving belt. It is the interplay of these opposing forces that permits the oscillations to be sustained.

If the block is initially given some nonzero velocity $\dot{x}$, the phase portrait shows that there could be one or more swings back and forth, each smaller than the previous one, until eventually the block adheres to the belt. Therefore, all orbits eventually tend to the cycle Γ except for the two equilibrium states and the smaller cyclic orbits which lie inside of Γ. However, these inner cycles are illusory because in actuality the transition from adhesion to slippage is continuous. For $\dot{x} - v \neq 0$, frictional force is better represented by the function shown in Figure 6.3 in which

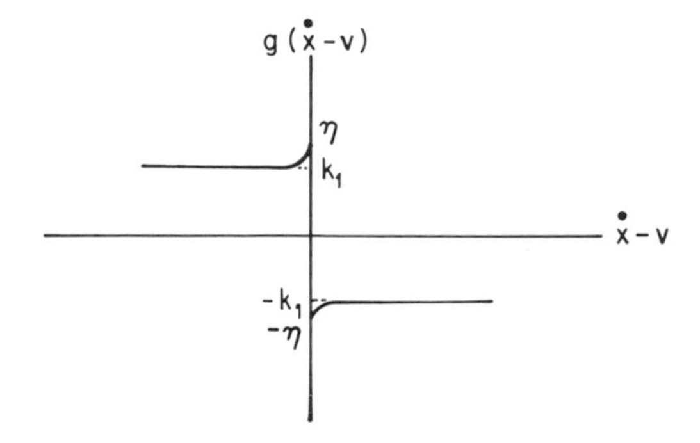

Figure 6.3 A more accurate portrayal of the frictional force $g(\dot{x} - v)$ for $\dot{x} - v \neq 0$. $g(\dot{x} - v)$ is roughly constant as before except for a slightly increasing slope when $\dot{x} < v$.

the slope of $g(\dot{x} - v)$ is positive (negative) near zero when $\dot{x} - v$ is negative (positive).

The rest point of the modified system is slightly displaced from its previous position, but if k_1 and η are not too far apart and if the slope of $g(\dot{x} - v)$ is small, then the rest point remains close to $\begin{pmatrix} \alpha/\omega^2 \\ 0 \end{pmatrix}$. The orbit structure also does not vary substantially from that shown in Figure 6.2. However, the equilibrium is now unstable, and the orbits spiral out from there (Exercise 6.5.1 and Figure 6.4). For this reason, Γ is called a limit cycle since all orbits, other than the stationary points, eventually approach Γ.

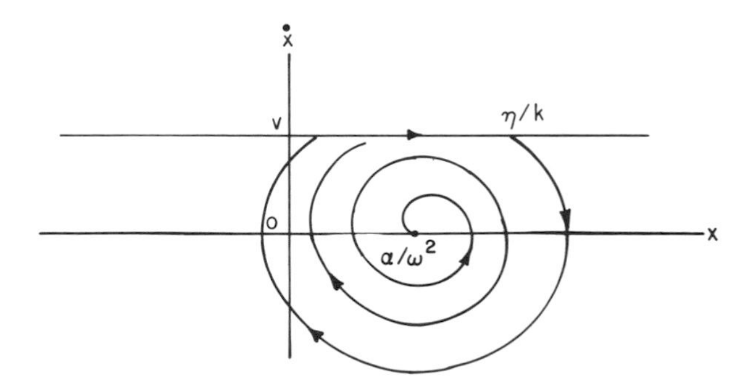

Figure 6.4 Orbit structure in the vicinity of the unstable equilibrium $x = \alpha/\omega^2$, $\dot{x} = 0$.

Consider now a stroked violin string which begins to vibrate. If the action of the bow is vigorous, represented by a large v, the amplitude of Γ is large as Figure 6.2 shows. This means a loud sound. However, the frequency of vibration of the string remains the same no matter how loud or soft it is stroked. It is easy to see why this must be so, at least for all v large enough, because Γ is very nearly an elliptical orbit in the x, $\dot{x}$ plane with a short straight line portion at $\dot{x} = v$. The elliptical part has simple harmonic motion of frequency ω (put $u = x - \alpha/\omega^2$ in 6.2 to obtain $\ddot{u} + \omega^2 u = 0$). The straight segment requires a traversal time of

$$T = \int_0^T dt = \int_{(2k_1 - \eta)/k}^{\eta/k} \frac{dx}{v},$$

which does depend on v, but is nevertheless negligible for v large. Thus, Γ has a period of roughly $2\pi/\omega = 2\pi m/k$, which depends on string mass and on the magnitude of the restoring force, namely the tension of the string.

Another example of a limit cycle is easily provided even though it has no immediate physical interpretation:

Example 6.2 Consider the system described by the equations

$$\begin{aligned} \dot{x}_1 &= x_2 + x_1(1 - x_1^2 - x_2^2) \\ \dot{x}_2 &= -x_1 + x_2(1 - x_1^2 - x_2^2). \end{aligned} \tag{6.4}$$

Let $x_1 = r\cos\theta$ and $x_2 = r\sin\theta$ for $r > 0$. Then 6.4 leads to

$$\begin{aligned} x_1\dot{x}_1 + x_2\dot{x}_2 &= r\dot{r} = r^2(1 - r^2) \\ x_2\dot{x}_1 - x_1\dot{x}_2 &= -r^2\dot{\theta} = r^2, \end{aligned}$$

which simplifies to

$$\begin{aligned} \dot{r} &= r(1 - r^2) \\ \dot{\theta} &= -1. \end{aligned} \tag{6.5}$$

The first equation in 6.5 is separable. If we let $u = 1/r$ in this equation and separate variables, then

$$\int \frac{u}{1 - u^2}\, du = -\frac{1}{2} \int \frac{d}{du} \ln(1 - u^2)\, du.$$

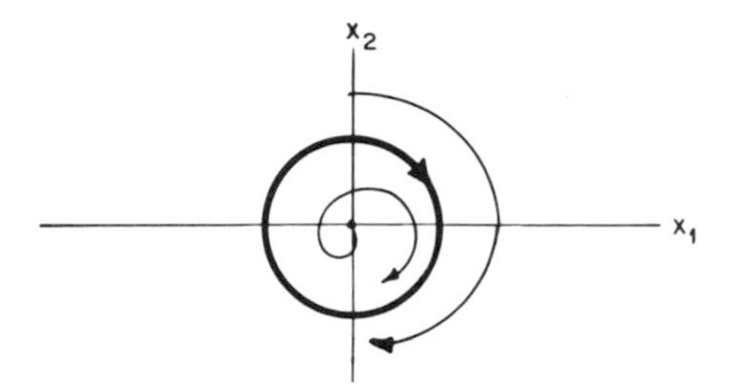

Figure 6.5 The unit circle as a limit cycle of Equations 6.4.

A straightforward integration gives

$$r(t) = \frac{r_0}{\sqrt{r_0^2 + (1 - r_0^2)e^{-2t}}}, \tag{6.6}$$

where $r_0 = r(0)$. The second equation in 6.5 shows that the radius vector r rotates clockwise at a constant speed. As $t \to \infty$ the orbit consequently spirals toward the unit circle in the x_1, x_2 plane no matter where it begins. That is, $r(t) \to 1$. It follows that circle of radius one is a limit cycle (Figure 6.5). The origin is the only equilibrium point of 6.5 and we conclude that it is unstable.

6.2 When Do Limit Cycles Exist?

We now know that limit cycles exist in at least some cases. Let us investigate this phenomenon a bit further by showing how such cycles can arise in general. To begin with, we need a definition. Denote an orbit of the system $\dot{\mathbf{x}} = \mathbf{f}(\mathbf{x}), \mathbf{x}$ in $\mathbf{R}^2$, by γ. The *positive limiting set* γ^+ is the set of points $\mathbf{p}$ in $\mathbf{R}^2$ for which

$$\|\mathbf{x}(t_n) - \mathbf{p}\| \to 0,$$

as $n \to \infty$, where $\{t_n\}$ is an increasing sequence of time. Two examples come immediately to mind. If γ is an equilibrium state, then it is apparent that γ^+ is identical with γ itself. Also when γ is a cycle, it is again true that $\gamma = \gamma^+$ because the orbit returns periodically to any $\mathbf{p}$ in γ.

We say that an orbit γ tends to a *limit cycle* Γ if $\gamma^+ = \Gamma$ and Γ is a nonconstant closed orbit.

A cycle γ is asymptotically stable if there is some open neighborhood Ω_0 of γ such that every orbit that begins in Ω_0 has γ as its positive limiting set. If no such Ω_0 can be found, then γ is unstable.

Recall (Section 3.1) that an invariant set Ω in the state space U has the property that if an orbit enters Ω it cannot later escape at some future time. The following result, a weak form of the *Poincaré–Bendixson Theorem* (consult the references for Chapter Six at the end of the book for details) guarantees limit cycles for two-dimensional dynamical systems (see also Figure 6.6).

Theorem 6.1 *Suppose Ω is a bounded invariant set in U for the system $\dot{\mathbf{x}} = \mathbf{f}(\mathbf{x})$, $\mathbf{x}$ in $\mathbf{R}^2$. Let γ be an orbit that begins in Ω. Then its positive limiting set γ^+ is nonempty, and if it contains no equilibria of the system, then either γ is a cycle or γ^+ is a limit cycle. On the other hand, if Ω contains no cycles but does contain a single attracting point $\mathbf{p}$, then $\gamma^+ = \mathbf{p}$.*

A typical application on the Poincaré–Bendixson Theorem is to the case in which there is a single unstable equilibrium $\bar{\mathbf{x}}$ in Ω that is repelling in the sense that all orbits in a neighborhood of $\bar{\mathbf{x}}$ move away from it. Simply remove this point. The remaining set of Ω sans point must then contain a limit cycle. In this case, it can also be shown that the cycle must surround the equilibrium. More generally, if the interior of a cycle γ in Ω belongs to U, then γ contains an equilibrium point. An illustration of this is found in Example 6.2 above: Let Ω be any disc about the origin in the plane whose radius is greater than unity. From Equations 6.5, we see that $\dot{r} < 0$ for $r > 1$ while $\dot{r} > 0$ when $r < 1$. It

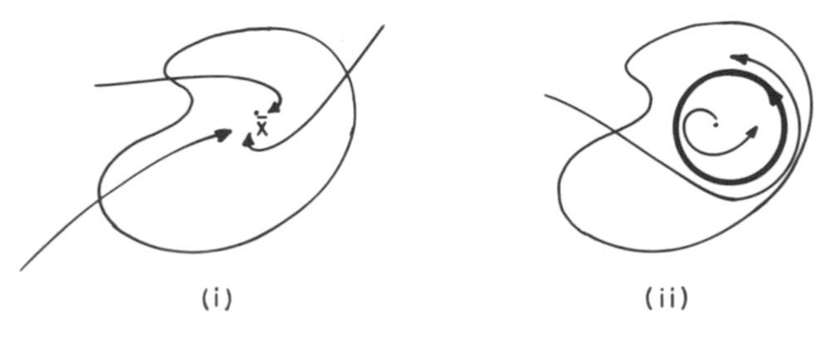

Figure 6.6 Illustration of Theorem 6.1. In (i) there is an isolated attracting equilibrium $\bar{\mathbf{x}}$ in an invariant set that contains no cycles while in (ii) the set contains a single repelling equilibrium. In this case, a limit cycle exists.

follows that Ω minus the origin is a bounded invariant set, and so a limit cycle exists. In this case, we know that γ^+ is the unit circle.

It is important to recognize that Theorem 6.1 is only valid in $\mathbf{R}^2$. The reason is that in the proof one uses the fact that cycles divide the plane into an inside and an outside. The uniqueness theorem does not permit orbits to cross from one of these regions to the other. In higher dimensions, this is no longer true and orbits can wind about each other in a complicated manner. This idea is taken up again in the next chapter.

Example 6.3 We are now in a position to complete the discussion begun in Section 5.5 of the previous chapter in which a traveling wave solution to Fisher's equation is being sought. The problem reduced to finding a function $\phi(s)$ that not only solves

$$\phi'' + r\phi(1 - \phi) + c\phi' = 0, \tag{6.7}$$

but also satisfies the conditions $0 \le \phi \le 1$, $\phi' < 0$, $\phi(-\infty) = 1$, and $\phi(\infty) = 0$. We saw that ϕ would have to be an orbit joining the unstable saddle at $\begin{pmatrix} 1 \\ 0 \end{pmatrix}$ to the stable point $\begin{pmatrix} 0 \\ 0 \end{pmatrix}$ and that the wave speed c must be large enough so that $c^2 \ge 4r$. To show that such a ϕ exists consider the triangular region Ω in the ϕ, ϕ' plane enclosed by the horizontal and vertical lines $\phi' = 0, \phi = 1$ as well as by the line l defined by $\phi' = -\mu\phi$ for μ positive. This line has an inward pointing normal given by the vector $\begin{pmatrix} \mu \\ 1 \end{pmatrix}$.

Writing 6.7 as a first-order system as in Section 5.5, we get:

$$\begin{aligned} u_1' &= u_2 \\ u_2' &= -cu_2 - ru_1(1 - u_1). \end{aligned}$$

The vector field defined by this system moves inward across the boundary of Ω provided that its inner product with the inward normal to Ω is positive. For example, along l the inner product is

$$\begin{pmatrix} \mu \\ 1 \end{pmatrix} \cdot \begin{pmatrix} u_2 \\ -cu_2 - ru_1(1 - u_1) \end{pmatrix} = \mu u_2 - cu_2 - ru_1(1 - u_1),$$

and since $u_2 = -\mu u_1$ along l, this is positive provided

$$\mu^2 - c\mu + r(1 - u_1) \le \mu^2 - c\mu + r < 0. \tag{6.8}$$

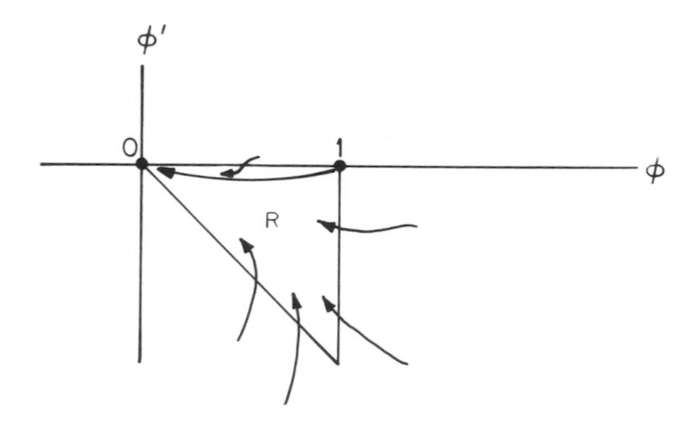

Figure 6.7 A triangular region in the ϕ, ϕ' phase plane contains an invariant region R for the flow of Equation 5.59. The orbit joining $\begin{pmatrix}1\\0\end{pmatrix}$ to $\begin{pmatrix}0\\0\end{pmatrix}$ satisfies $0 \le \phi \le 1$, $\phi' < 0$, $\phi(-\infty) = 1$, and $\phi(\infty) = 0$.

The quadratic $\mu^2 - c\mu + r$ has two positive real roots since $c^2 \ge 4r$, and so 6.8 is valid for some range of positive μ. A similar argument applies to each of the other bounding edges of Ω. Hence, Ω is invariant to the flow of 6.7 since all orbits bend inward along the boundary. There are no cycles in Ω for they would have to enclose a rest point. There are only two of these, and they are on the boundary. Since $\begin{pmatrix}1\\0\end{pmatrix}$ is repelling within Ω, its unstable manifold must tend to the attractor at the origin. Thus, $\phi(-\infty) = 1$ and $\phi(\infty) = 0$ as required. This application of the Poincaré–Bendixson Theorem is illustrated in Figure 6.7.

There is a second result that does apply to dynamical systems of higher dimension than two. However, we discuss it first for planar systems and then state the three-dimensional version later.

In a number of models, the equations that describe the dynamics include a parameter μ that is allowed to vary. In $\mathbf{R}^2$ such equations may be written as

$$\begin{aligned}\dot{x}_1 &= f_1(x_1, x_2, \mu)\\ \dot{x}_2 &= f_2(x_1, x_2, \mu)\end{aligned} \tag{6.9}$$

or, in vector notation, as

$$\dot{\mathbf{x}} = \mathbf{f}(\mathbf{x}, \mu). \tag{6.10}$$

For technical reasons, the components of the vector field $\mathbf{f}$ are assumed to be polynomials in $\mathbf{x}$ and μ. This restriction is satisfied by all the examples in this book. For each μ in some range, suppose there is an isolated equilibrium $\bar{\mathbf{x}} = \begin{pmatrix} \bar{x}_1 \\ \bar{x}_2 \end{pmatrix}$ that depends, in general, on the choice of μ. Indicate this dependence by writing $\bar{\mathbf{x}}(\mu)$. The linearized system corresponding to 6.9 has a Jacobian matrix A that may also depend on μ. That is,

$$A(\mu) = \left. \begin{pmatrix} \dfrac{\partial f_i}{\partial x_1} & \dfrac{\partial f_1}{\partial x_2} \\ \dfrac{\partial f_2}{\partial x_1} & \dfrac{\partial f_2}{\partial x_2} \end{pmatrix} \right|_{\bar{\mathbf{x}}(\mu)} . \tag{6.11}$$

The eigenvalues of $A(\mu)$ are given by $\lambda_i(\mu)$, $i = 1, 2$. We suppose that for some suitable range of $|\mu|$ values, say $|\mu| < \delta$, the eigenvalues are differentiable in μ and complex:

$$\lambda_i(\mu) = \alpha(\mu) \pm i\beta(\mu).$$

The following result is known as the *Hopf Bifurcation Theorem* in $\mathbf{R}^2$.

Theorem 6.2 *Suppose that the equilibrium $\bar{\mathbf{x}}(\mu)$ is a.s. for $\mu < 0$ and unstable for $\mu > 0$ and that $\alpha(0) = 0$. If $\dfrac{d\alpha(0)}{d\mu} > 0$ and $\beta(0) \neq 0$ then for all sufficiently small $|\mu|$, a closed orbit exists for μ either positive or negative. In particular, if $\bar{\mathbf{x}}(0)$ is locally a.s., then there is a stable limit cycle Γ about $\bar{\mathbf{x}}(\mu)$ for all small $\mu > 0$. Moreover, the amplitude of Γ grows as μ increases.*

An intuitive argument makes the theorem plausible, at least in the case in which $\bar{\mathbf{x}}(0)$ is a.s. in some basin of attraction Ω_0. The Liapunov Theorem (Theorem 3.1) guarantees the a.s. of $\bar{\mathbf{x}}(0)$ if there is some Liapunov function V on Ω_0 with $\dot{V} < 0$ everywhere except at $\bar{\mathbf{x}}$. It is entirely reasonable to suppose that the converse also holds: V must exist since $\bar{\mathbf{x}}$ is a.s. on Ω_0. In fact, this may be shown to be true under reasonable conditions, although we do not do so here. The orbits of 6.10 cross the level of curves of V in an inward direction on Ω_0. If μ increases slightly to some positive value, the orbits will also shift slightly since 6.10 is smooth in $\mathbf{x}$ and μ. Therefore, inward orbits remain inward in direction for small enough μ. If the repelling unstable equilibrium $\bar{\mathbf{x}}(\mu)$ is removed from Ω_0, the remaining set is bounded and invariant, and so the Poincaré–Bendixson Theorem assures us of a limit cycle about $\bar{\mathbf{x}}(\mu)$ for each such μ. A formal proof requires more care and is less straightforward.

In Theorem 6.2, the real parts of the eigenvalues of $A(\mu)$ cross the imaginary axis as μ moves past the origin $\mu = 0$ from left to right (since $\alpha'(0) > 0$). In general, it is not readily determined whether or not the equilibrium $\bar{\mathbf{x}}$ is locally a.s. at $\mu = 0$ since $\alpha(0) = 0$ means that $\bar{\mathbf{x}}$ is nonhyperbolic, which of course precludes any deduction from a knowledge of the linearized system alone. The linearized system is neutrally stable about $\bar{\mathbf{x}}(0)$ since the eigenvalues are imaginary. In Example 6.4, which follows, a suitable Liapunov function overcomes this obstacle, but in our other examples we are left in the dark. The Hopf Theorem is therefore somewhat ambiguous about the nature of the cycle. One possibility is that an unstable cycle exists for $\mu < 0$, inside of which all orbits spiral in toward $\bar{\mathbf{x}}(\mu)$. Another is that orbits spiral outward from $\bar{\mathbf{x}}(\mu)$ when $\mu > 0$ toward a stable limit cycle (Figure 6.8). Note that the

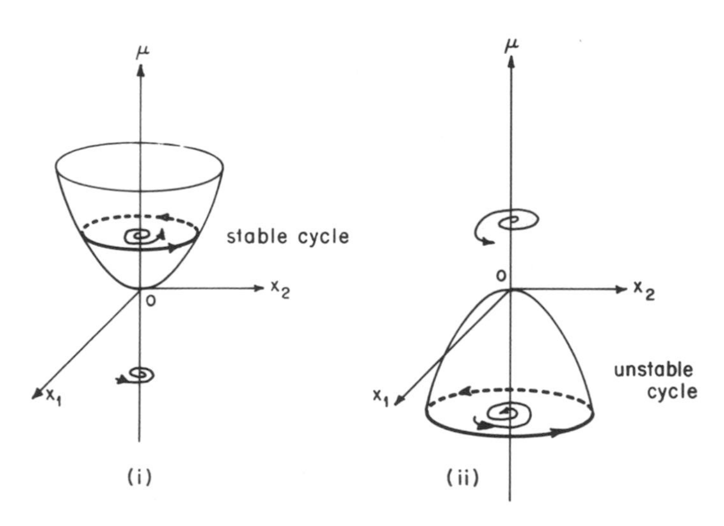

Figure 6.8 Pictorial representation of Theorem 6.2 in terms of x_1, x_2 and μ. In (i) a stable limit cycle exists for $\mu > 0$, while in (ii) an unstable limit cycle reigns for $\mu < 0$. In both cases $\bar{\mathbf{x}}(\mu) = 0$ for all μ.

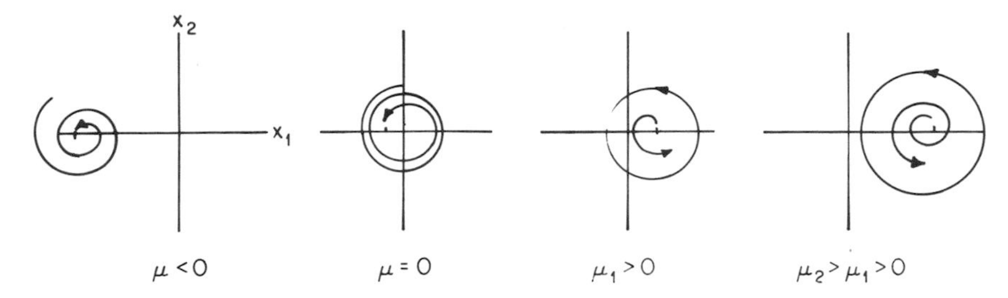

Figure 6.9 Another illustration of Theorem 6.2: $\bar{\mathbf{x}}(\mu)$ is a.s. for $\mu < 0$. At $\mu = 0$ stability is maintained, but just barely. As μ increases beyond zero a stable periodic orbit grows. $\bar{\mathbf{x}}(\mu)$ is shown moving to the right as μ increases although this need not be the case in general.

unstable cycle is a limit cycle if time is run backward ($t \to -\infty$). There is yet another possibility, which is that all cycles occur only at $\mu = 0$ (Exercise 6.5.3). For $\mu = 0$, the cycles degenerate in this case to the equilibrium point itself.

The critical value of zero for the parameter μ is called a *Hopf bifurcation point*. We have seen, in summary, that a periodic orbit is either expelled or absorbed at $\mu = 0$. In one case, local instability gives way to a globally stable closed orbit, and in the other, a globally unstable cycle degenerates into a stable point. The first instance is illustrated again in the Figure 6.9. It is useful to point out that $\bar{\mathbf{x}}(\mu)$ can always be chosen as zero for all μ simply by letting $\mathbf{z} = \mathbf{x} - \bar{\mathbf{x}}(\mu)$ and rescaling 6.10 to $\dot{\mathbf{z}} = \mathbf{g}(\mathbf{z}, \mu)$ for some function $\mathbf{g}$. The straightforward details are left to the reader (see, however, Exercise 1.5.5).

Example 6.4 Consider a modified version of the Van der Pol equation, which was considered in Example 2.5. In new notation the equations are

$$\begin{aligned} \dot{x}_1 &= -x_2 \\ \dot{x}_2 &= x_1 - \alpha\left(\frac{x_2^3}{3} - \mu x_2\right), \qquad \alpha > 0. \end{aligned} \tag{6.12}$$

Note the addition of a parameter μ. The equilibrium $\bar{\mathbf{x}}$ is in all cases zero, independent of μ, and the Jacobian matrix of the linearized system is easily computed to be

$$A(\mu) = \begin{pmatrix} 0 & -1 \\ 1 & \alpha(\mu - x_2^2) \end{pmatrix}\Bigg|_{\bar{\mathbf{x}}=\mathbf{0}} = \begin{pmatrix} 0 & -1 \\ 1 & \alpha\mu \end{pmatrix}.$$

Det $A(\mu) > 0$ for all μ, but Trace $A(\mu) = \alpha\mu$ depends on the sign of μ. For $\mu < 0$, the equilibrium state is a.s. and it is unstable for $\mu > 0$. At $\mu = 0$, the eigenvalues are imaginary and no judgment about stability can be made. However, if we form the function

$$V(x_1, x_2) = \frac{x_1^2 + x_2^2}{2},$$

then, along orbits of 6.12, one has

$$\dot{V} = x_1\dot{x}_1 + x_2\dot{x}_2 = -\alpha x_2^2\left(\frac{x_2^2}{3} - \mu\right).$$

When $\mu = 0$, $\dot{V}$ is negative except for $x_2 = 0$. However, the origin is the only invariant subset of the set in which $\dot{V} = 0$, and so V is a Liapunov function (Theorem 3.1). Hence, $\bar{\mathbf{x}} = \mathbf{0}$ is a.s.

The eigenvalues of $A(\mu)$ are complex for $|\mu|$ small enough with $\text{Re}\,\lambda_i(\mu) = \alpha\mu/2$ satisfying

$$\frac{d\,\text{Re}\,\lambda_i(\mu)}{d\mu} = \frac{\alpha}{2} > 0$$

and

$$\text{Re}\,\lambda_i(0) = 0.$$

Also,

$$\text{Im}\,\lambda_i(0) \neq 0.$$

It follows from Theorem 6.2 that a stable limit cycle exists about the origin for all $\mu > 0$ small enough. Just how small μ must be is not revealed by Theorem 6.2. However, it is known that a stable periodic orbit does in fact exist for $0 < \mu \leq 1$ ($\mu = 1$ is the original Van der Pol equation) by using Theorem 6.1 directly albeit in a fairly complicated way. We saw in Example 3.5 that the origin for the Van der Pol equation is unstable when $\alpha > 0$. We now know that the orbits repelled by the origin tend to a limit cycle.

The Van der Pol equation can be interpreted as a mass–spring system in which the frictional force has the curious property that although energy is dissipated for $\|\mathbf{x}\|$ large, it is pumped into the system for $\|\mathbf{x}\|$

small (Example 2.6). It is the interplay between the destabilizing effect of excitation for small $\|\mathbf{x}\|$ and the damping that takes place for large $\|\mathbf{x}\|$ that gives rise to self-sustained oscillations.

There is a version of Theorem 6.2 that is valid in $\mathbf{R}^k$. Consider the case $k = 3$. As before, $\mathbf{f}(\mathbf{x}, \mu)$ is a polynomial in $\mathbf{x}$ and μ. The system $\dot{\mathbf{x}} = \mathbf{f}(\mathbf{x}, \mu)$, $\mathbf{x}$ in $\mathbf{R}^3$, can be linearized about an isolated equilibrium $\bar{\mathbf{x}}(\mu)$ by Taylor's Theorem, exactly as in Theorem 2.1. However, the Jacobian is now a 3×3 matrix

$$A(\mu) = \left. \begin{pmatrix} \dfrac{\partial f_1}{\partial x_1} & \dfrac{\partial f_1}{\partial x_2} & \dfrac{\partial f_1}{\partial x_3} \\ \dfrac{\partial f_2}{\partial x_1} & \dfrac{\partial f_2}{\partial x_2} & \dfrac{\partial f_2}{\partial x_3} \\ \dfrac{\partial f_3}{\partial x_1} & \dfrac{\partial f_3}{\partial x_2} & \dfrac{\partial f_3}{\partial x_3} \end{pmatrix} \right|_{\bar{\mathbf{x}}(\mu)} \tag{6.13}$$

There are three eigenvalues $\lambda_i(\mu)$, and we suppose that two of them are complex, say λ_1 and λ_2, while λ_3 is real (and negative at $\mu = 0$). All λ_i are differentiable in μ. Then Theorem 6.2 goes through word for word. The Hopf Theorem in $\mathbf{R}^3$ will be used later.

While we are at it, note that the linearized equations corresponding to $\dot{\mathbf{x}} = \mathbf{f}(\mathbf{x}, \mu)$ for any given μ, $\mathbf{x}$ in $\mathbf{R}^3$, can be written as

$$\dot{\mathbf{u}} = A(\mu)\,\mathbf{u} \tag{6.14}$$

with $\mathbf{u} = \mathbf{x} - \bar{\mathbf{x}}$ a vector in $\mathbf{R}^3$. The local stability or instability of the nonlinear system proceeds exactly as before (Theorem 3.1) provided the eigenvalues $\lambda_i(\mu)$ of $A(\mu)$ are hyperbolic (namely $\operatorname{Re}\lambda_i(\mu) \neq 0$). From the solution of 6.14 (see the Appendix for a discussion of equations in $\mathbf{R}^3$) it is easy to establish, just as in Theorem 2.1, that the origin is a.s. if and only if $\operatorname{Re}\lambda_i(\mu) < 0$. That is why we insist that λ_3 be negative in the vicinity of $\mu = 0$.

The word bifurcation connotes a parting of the ways. In the Hopf bifurcation, an equilibrium state either absorbs or spews out a closed orbit: local stability or instability is exchanged for the global opposite at the critical value $\mu = 0$. There are other kinds of bifurcation, however. In one of these, the *pitchfork bifurcation*, a stable equilibrium splits into two other stable equilibria as a parameter μ passes a critical value. Although

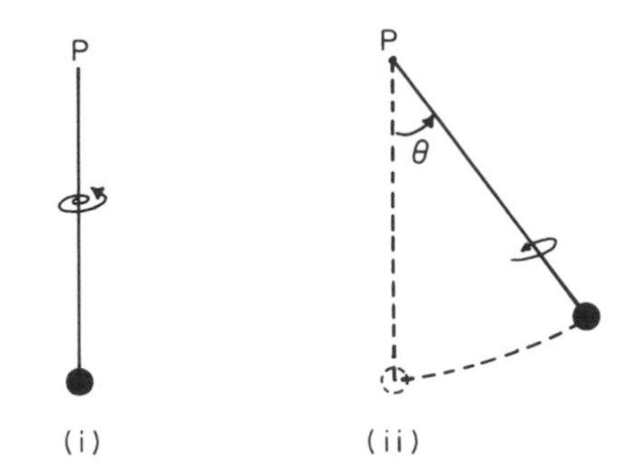

Figure 6.10 A pendulum is rotated at speed ω around a vertical axis of length l about a movable joint P. In (i) $\omega < \omega_0$ and the rod remains vertical as it rotates, while in (ii) the rod begins to rise as ω increases beyond ω_0.

this situation is taken up again in Chapter Seven, we want to illustrate it here since the computation in the next example will be used anew in a later section.

Example 6.5 Consider again the pendulum (Section 2.1). A bob of mass m is attached to a massless rigid rod of length l, and the rod is hinged to a movable joint. This time, however, instead of displacing the bob from rest and letting it swing back and forth in a planar motion, it is simply allowed to hang freely. The rod is then rotated at angular speed ω about its pivot in a plane perpendicular to the page. It is noticed that as ω increases a critical speed ω_0 is reached at which the rod begins to rise up as it rotates. The bob and rod configuration is then displaced by an angle θ from the vertical (Figure 6.10), and it sweeps out a conical surface as it rotates.

There are two main forces at work here. One is the gravitational force mg on the mass pulling it downward. The other is the horizontal force due to the centripetal acceleration resulting from rotational motion. The rod exerts a restraining force T, whose horizontal and vertical components are $T \sin\theta$ and $T\cos\theta$ respectively. If the pendulum is to rotate at a constant angle θ, then the vertical forces must balance each other in magnitude:

$$T\cos\theta = mg.$$

However, an upward movement of the bob requires unbalanced forces:

$$T\cos\theta > mg. \tag{6.15}$$

Unless 6.15 is satisfied, the rod remains straight down and θ is zero. The distance of the bob from the center of rotation is $l \sin \theta$. Centripetal acceleration is then $l\omega^2 \sin \theta$, and so the horizontal component of the restraining force T must satisfy

$$T \sin \theta = ml\omega^2 \sin \theta. \tag{6.16}$$

Combining 6.15 and 6.16 gives

$$\cos \theta > \frac{g}{l\omega^2}. \tag{6.17}$$

If $g/l\omega^2 \geq 1$, then 6.17 is not satisfied by any θ, and so the bob rotates in a vertical position. As angular speed ω increases, a certain critical ω_0 is reached beyond which 6.15 can be met. This occurs when

$$\frac{g}{l\omega_0^2} = 1 \qquad \text{or} \qquad \omega_0 = \sqrt{\frac{g}{l}}.$$

For $\omega > \omega_0$, 6.17 holds with $\theta > 0$. Letting $\mu = \omega - \omega_0$, we see that a pitchfork bifurcation takes place at $\mu = 0$. A zero angle of displacement splits into one of the possible angles $\pm\theta$ depending on which side the bob first begins to rise.

It is worthwhile to derive the dynamical equation for θ since the result is needed later anyhow (Section 6.4). To this end, reconsider the force diagram on the bob as it moves along a circular arc a distance $s = l\theta$ by taking the components of the forces $T \cos \theta$ and mg tangentially along the arc (Figure 6.11). Acceleration along the arc is $\ddot{s} = l\ddot{\theta}$. This is determined by the difference in the forces moving the bob either up or down:

$$m\ddot{s} = ml\ddot{\theta} = T \cos \theta \sin \theta - mg \sin \theta. \tag{6.18}$$

But $T = ml\omega^2$ from 6.16, and so 6.18 becomes

$$\ddot{\theta} = \omega^2 \cos \theta \sin \theta - \frac{g}{l} \sin \theta. \tag{6.19}$$

Letting $x_1 = \theta$ and $x_2 = \dot{\theta}$, we can rewrite 6.19 as a first-order system:

$$\begin{aligned} \dot{x}_1 &= x_2 \\ \dot{x}_2 &= \omega^2 \cos x_1 \sin x_1 - \frac{g}{l} \sin x_1. \end{aligned} \tag{6.20}$$

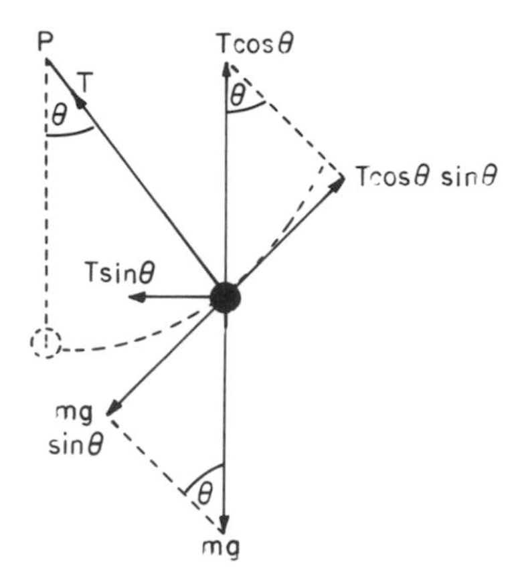

Figure 6.11 Force diagram for the rotating pendulum. The horizontal and vertical forces $T\cos\theta$ and mg are resolved along the tangent to the arc.

If $g/l\omega^2 \geq 1$, the only equilibrium is at $\bar{\mathbf{x}} = \begin{pmatrix} 0 \\ 0 \end{pmatrix}$. A linearized stability analysis shows it to be stable for $g/l\omega^2 > 1$, that is for

$$\omega < \omega_0 = \sqrt{\frac{g}{l}},$$

and unstable for $\omega > \omega_0$. When $\omega > \omega_0$, there are additional equilibria determined by

$$\cos x_1 = \frac{g}{l\omega^2} \quad \text{and} \quad x_2 = 0. \tag{6.21}$$

Relation 6.21 defines two angles $\theta \equiv x_1 = \arccos(g/l\omega^2)$ depending on which side of the vertical the rod begins to rise. Another stability check shows these equilibria to be stable for any given $\omega > \omega_0$. Thus, a stable position of θ for $\mu = \omega - \omega_0 < 0$ bifurcates at $\mu = 0$ into two stable angles $\pm\theta$ for $\mu > 0$. The original position $\theta = 0$ becomes unstable for $\mu > 0$.

Let us note for later use that if frictional resistance of the pivot at P induces a viscous damping force to counteract motion, then 6.18 must be modified by the inclusion of the tangential component of this force which is proportional to the velocity $\dot{s} = l\dot{\theta}$ along the arc. Therefore, 6.19

is augmented to become

$$\ddot{\theta} = \omega^2 \cos\theta \sin\theta - \frac{g}{l} \sin\theta - r\dot{\theta}, \tag{6.22}$$

where $r > 0$ is an appropriate constant. It can be shown that the equilibrium $\theta = \dot{\theta} = 0$ of 6.21 is globally a.s. for $\omega < \sqrt{\frac{g}{l}}$ (Exercise 6.5.5).

6.3 The Struggle for Life, II

In order to illustrate the concept of a Hopf bifurcation let us turn to two examples which are based on the population growth models of Chapter Four.

Example 6.6 (Predator–Prey) The quadratic predator–prey model encoded in Equations 4.22 has the property that the nontrivial equilibrium in the positive quadrant is globally a.s. The exception to this is the neutrally stable case of 4.21 in which there are a family of closed orbits. It appears that cyclic behavior is not uncommon in certain two species interactions, and so the closed orbits could be revealing until we recall that this neutrally stable case is not robust in the sense that even slightest change in the equation structure alters the resulting phase portrait drastically. A perhaps more satisfactory model should take into account the idea that even though the predator eats as much as it can find when food is scarce, it is not unreasonable that during periods of abundance the predator satiates, and it then feeds at a maximum possible per-capita rate β, independent of the number of available prey. A useful way of expressing this is to consider a predation term of the form

$$\frac{\beta N_1 N_2}{\alpha + N_1} \tag{6.23}$$

in which α is some positive constant. When N_1 is small, 6.23 is proportional to $N_1 N_2$ as in 4.21. However, for N_1 large this expression is roughly βN_2, as required. The predator is assumed to grow logistically but with a carrying capacity that is proportional to the amount of food available. Letting N_1, N_2 be expressed as x_1, x_2, these conditions permit a modified version of 4.21, one of several possible:

$$\begin{aligned}\dot{x}_1 &= rx_1\left(1 - \frac{x_1}{K}\right) - \frac{\beta x_1 x_2}{\alpha + x_1}\\ \dot{x}_2 &= sx_2\left(1 - \frac{x_2}{\nu x_1}\right)\end{aligned} \tag{6.24}$$

in which s and ν are positive constants. This model, due to R. May, admits a stable limit cycle about an unstable equilibrium, as we now wish to show.

Ignoring the extinction equilibrium in which $\bar{x}_1 = K$, $\bar{x}_2 = 0$, there is only one nontrivial rest state defined by

$$\begin{aligned}\bar{x}_2 &= \nu\bar{x}_1\\ r\left(1 - \frac{\bar{x}_1}{K}\right) &- \frac{\beta\bar{x}_2}{\alpha + \bar{x}_1} = 0.\end{aligned} \tag{6.25}$$

Observe that $\bar{x}_1$, and hence $\bar{x}_2$, depends on all the parameters except s.

The usual isocline analysis suggests the orbit structure in Figure 6.12, in which there appears to be a cycle.

Let $\mu = \bar{s} - s$, where $\bar{s} = -r\bar{x}_1/K + \beta\nu\bar{x}_1^2/(\alpha + \bar{x}_1)^2$ is a fixed quantity independent of s. The Jacobian matrix of the linearized system evaluated at $\bar{\mathbf{x}}$ can be written, by virtue of 6.24, as

$$A(\mu) = \begin{pmatrix} \bar{s} & -\beta\bar{x}_1/(\alpha + \bar{x}_1) \\ (\bar{s} - \mu)\nu & \mu - \bar{s} \end{pmatrix}.$$

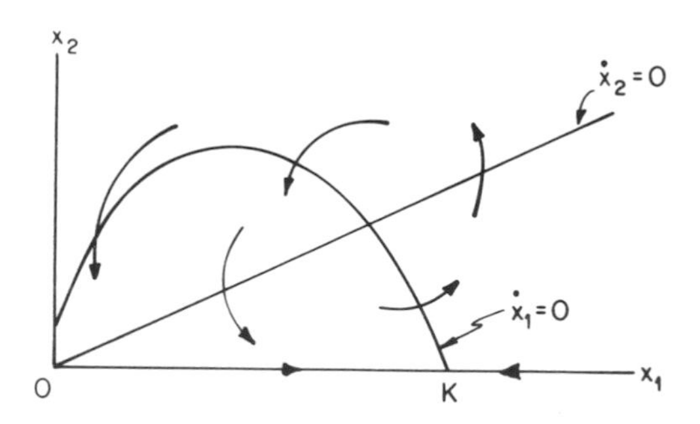

Figure 6.12 Isoclines for Equation 6.24 together with an anticipated orbit structure.

It follows that Det $A(\mu)$ is positive, independent of s and Trace $A(\mu) = \mu$. Therefore, the eigenvalues $\lambda_i(\mu)$ of $A(\mu)$ are complex for all $|\mu|$ small enough. Also, $\operatorname{Re}\lambda_i(0) = 0$ if and only if $\mu = 0$ since a zero trace is equivalent to imaginary eigenvalues. Therefore, $\operatorname{Im}\lambda_i(0) \neq 0$ and

$$\frac{d\operatorname{Re}\lambda_i(\mu)}{d\mu} = \frac{1}{2}\frac{d\operatorname{Trace}A(\mu)}{d\mu} = \frac{1}{2} > 0.$$

Note that $\bar{\mathbf{x}}(\mu)$ is a.s. for $\mu < 0$ and unstable when $\mu > 0$. The case $\mu = 0$ is undecided.

By Theorem 6.2 a limit cycle exists for all $|\mu|$ small enough. The isoclines in Figure 6.12 actually imply a stable cycle for $\mu > 0$ in certain cases because of the Poincaré–Bendixson Theorem. Indeed, by following any orbit that begins near $\begin{pmatrix} K \\ 0 \end{pmatrix}$ in the region where $\dot{x}_1 < 0$, we see that the orbit is trapped in a bounded invariant set and that it cycles about the equilibrium $\bar{\mathbf{x}}$ (where the isoclines intersect). In fact, it cannot tend toward the vertical axis ($\dot{x}_2 \to -\infty$ as $\dot{x}_1 \to 0$) nor toward the unstable point $\begin{pmatrix} K \\ 0 \end{pmatrix}$, and since $\bar{\mathbf{x}}$ is repelling, a limit cycle exists (Figure 6.13). The same reasoning applies to any orbit other than the equilibrium states.

Note that μ is always negative if $\bar{s} < 0$. However, if K gets arbitrarily large relative to the equilibrium value $\bar{x}_1$, then $\bar{s}$ is eventually positive and, with s small enough, μ is also positive.

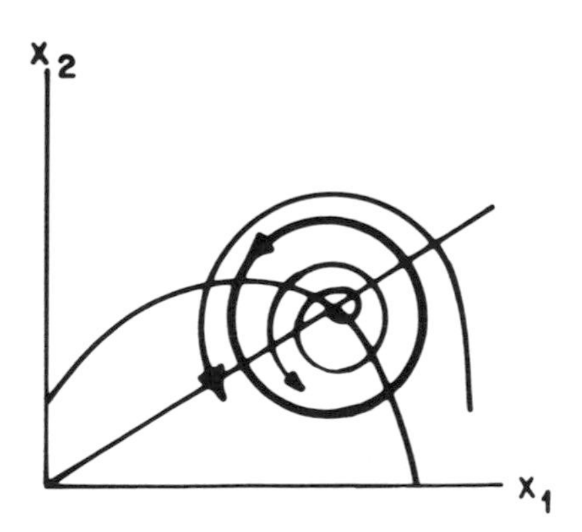

Figure 6.13 The orbit structure of 6.24 when $\mu > 0$. The predator and prey coexist as they spiral towards a limit cycle. If $\mu < 0$ there is also the possibility of the two species settling down to a stable equilibrium.

A small s relative to r suggests that the prey may be only mildly affected by predation. Suppose, in addition, that K is large. This implies that the prey has available an abundant supply of food. In this situation, the possibility of a limit cycle means that the prey population can fluctuate cyclically to levels below that which it would have if $\mu < 0$. Paradoxically, this says that a sufficiently enriched habitat can lead to observed population levels of prey which are lower than would occur under leaner, more austere, circumstances.

Example 6.7 (Fish Harvesting) Equation 4.7 models a species of fish growing logistically but harvested at a constant per-capita rate νE. This is rewritten as:

$$\dot{N} = rN\left(1 - \frac{N}{K}\right) - \nu EN \tag{6.26}$$

in which the "catchability coefficient" ν measures the fact that the same effort E, in terms of fleet and equipment size, can yield different catches depending on how efficient the fleet is and on difficulties of harvest which vary with locale.

Instead of keeping E constant, we now allow it to vary over time in response to the economics of fishing. Specifically, suppose that p is the per unit price of fish on the open market. Since νEN is the harvest rate, the revenue from the harvest is νpEN. E is a measure of the effort expended in operating a fishing fleet, and we take the operating cost to be proportional to E, viz cE. Both p and c are positive constants even though in general they can vary (this model is reconsidered in Chapter Nine). Hence, the net income from fishing is $E(\nu pN - c)$. If this quantity is positive, it is reasonable to suppose that fishing effort would increase with just the opposite being true in the reverse situation. Therefore, the change in E due to market conditions is modeled as

$$\dot{E} = \alpha E(\nu pN - c), \tag{6.27}$$

where $\alpha > 0$ is some constant of proportionality. As it stands, the dynamical system consisting of 6.26 and 6.27 has a globally a.s. equilibrium (Exercise 6.5.6).

In order to make the model more interesting, we replace the logistic growth curve by one in which the per-capita growth rate increases for N small enough. A typical situation is illustrated in Figure 4.2. In the language of Section 5.6 such growth is called autocatalytic, at least for small N. A concrete example is given by

$$f(N) = N\left(r + \frac{(\beta - N)N}{1 + \alpha N}\right) \tag{6.28}$$

with r, α, and β all positive. The per-capita growth rate $f(N)/N$ has a positive slope for all N less than β. Thereafter, $f(N)/N$ decreases until N reaches some equilibrium value $\overline{N}$, where it is then zero. The usual argument shows that $\overline{N}$ is a.s. for all $N \geq 0$. Observe that for N large enough, 6.15 can be approximated by $N(r + \beta/\alpha - N/\alpha)$, which is roughly of logistic form. In what follows, it is not necessary to assume that $f(N)$ has the specific form of 6.28 although one may wish to do so for the sake of concreteness. All that we assume is $f(N)/N \geq 0$ for $0 < N < \overline{N}$ and

$$\frac{d}{dN}\left(\frac{f(N)}{N}\right) > 0 \qquad \text{for } N < \hat{N}, \tag{6.29}$$

where $\hat{N}$ is some number less than $\overline{N}$.

The fishery model is therefore

$$\begin{aligned} \dot{N} &= f(N) - \nu EN \\ \dot{E} &= \alpha E(\nu pN - c). \end{aligned} \tag{6.30}$$

There is a nontrivial equilibrium at $\overline{N} = c/\nu p$, $\overline{E} = f(\overline{N})/\nu\overline{N}$. Letting $\mu = \hat{N} - \overline{N}$, the Jacobian matrix of the linearized system at the equilibrium is

$$A(\mu) = \begin{pmatrix} f'(\hat{N} - \mu) - \nu\overline{E} & -\dfrac{c}{p} \\ \nu\alpha p\overline{E} & 0 \end{pmatrix}.$$

Therefore, Det $A(\mu) > 0$ for all μ, while

$$\operatorname{Trace} A(\mu) = f'(\overline{N}) - \frac{f(\overline{N})}{\overline{N}}. \tag{6.31}$$

To decide on the sign of the trace consider any straight line emanating from the origin in the N, $f(N)$ plane and intersecting $f(N)$. Its slope is

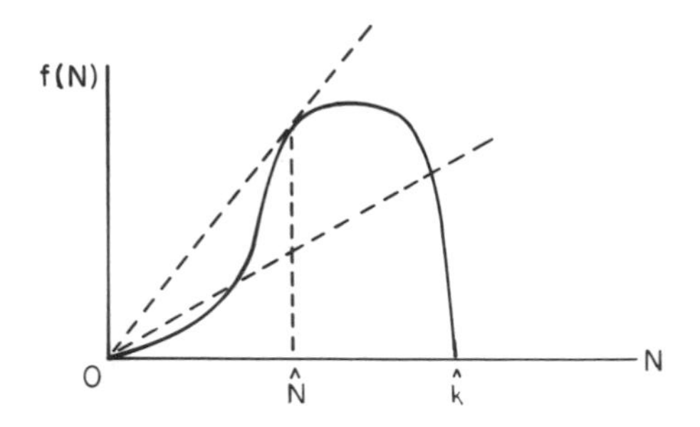

Figure 6.14 Two straight lines (dotted) which intersect a growth curve $f(N)$ of the form 6.28.

$f(N)/N$ (Figure 6.14). It is tangent to $f(N)$ at $\hat{N}$ but for $N < \hat{N}$ $(N > \hat{N})$ the slope is less than (greater than) $f'(N)$.

Therefore, since $\overline{E} = f(\overline{N})/\nu\overline{N}$,

$$\begin{aligned} \overline{E} &< f'(\overline{N}) \qquad \text{if } \overline{N} < \hat{N}, \\ \overline{E} &= f'(\hat{N}) \qquad \text{if } \overline{N} = \hat{N}, \\ \overline{E} &> f'(\overline{N}) \qquad \text{if } \overline{N} > \hat{N}. \end{aligned}$$

It follows from 6.31 that $\operatorname{Trace} A(\mu) < 0$ if and only if $\overline{N} > \hat{N}$, or equivalently if $\mu < 0$. Moreover, $\operatorname{Trace} A(\mu) = 0$ only if $\mu = 0$, and so the eigenvalues $\lambda_i(\mu)$ of $A(\mu)$ satisfy

$$\operatorname{Re} \lambda_i(\mu) < 0 \qquad \text{for } \mu < 0,$$

and $\operatorname{Re} \lambda_i(0) = 0$. When $\mu > 0$, there is instability. It is easy to see that $\lambda_i(\mu)$ are complex for $|\mu|$ small and that $\operatorname{Im} \lambda_i(0) \neq 0$. Finally,

$$\begin{aligned} \frac{d \operatorname{Re} \lambda_i(\mu)}{d\mu} &= \frac{1}{2} \frac{d \operatorname{Trace} A(\mu)}{d\mu} \\ &= \frac{1}{2}\left[f''(\overline{N}) \frac{d\overline{N}}{dN} - \frac{d}{d\overline{N}}\left(\frac{f(\overline{N})}{\overline{N}}\right)\frac{d\overline{N}}{d\mu}\right] \\ &= -\frac{1}{2}\left[f''(\overline{N}) + \frac{f'(\overline{N})}{\overline{N}} - \frac{f(\overline{N})}{\overline{N}^2}\right]. \end{aligned}$$

From this relation one sees that

$$\frac{d \operatorname{Re} \lambda_i(0)}{d\mu} = -\frac{1}{2} f''(\overline{N}) > 0$$

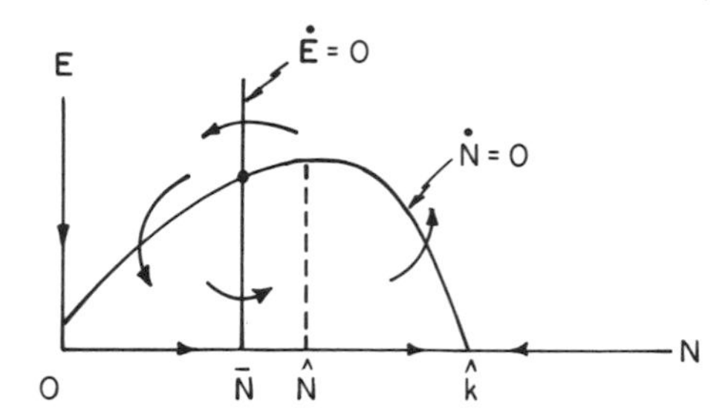

Figure 6.15 Isocline analysis of Equation 6.30.

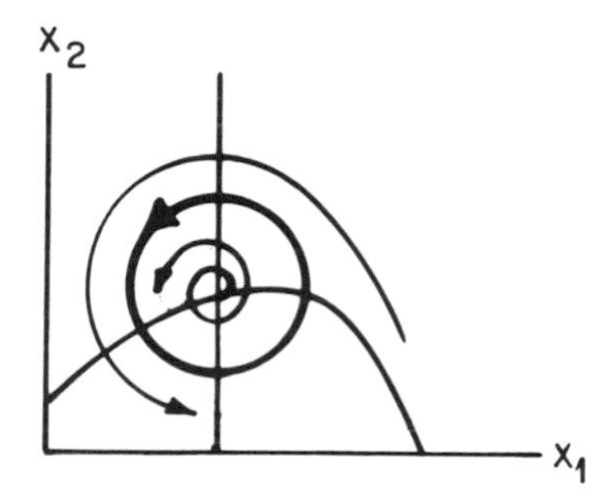

Figure 6.16 The orbit structure of 6.30 when $\mu > 0$. The fishing effort fluctuates cyclically in response to a cyclic change in the level of fish.

since $f'(\hat{N}) = f(\hat{N})/\hat{N}$ (at $\mu = 0$) and $f''(\hat{N}) < 0$. This last inequality follows from the fact that $f(N)/N$ reaches a maximum at $\hat{N}$ (Exercise 6.5.7).

The conditions of the Hopf bifurcation theorem are satisfied, and so a limit cycle exists for $|\mu|$ small enough. Whether this is a stable cycle for $\mu > 0$ is not clear, but we can argue as in the preceding example. An isocline analysis reveals, as before, that if $\mu > 0$ (in which case the equilibrium is unstable), then all orbits either spiral towards a cycle or they tend to infinity along the vertical axis where $N = \hat{N}$. However, the second equation in 6.30 shows that $\dfrac{d \ln E}{dt} \to 0$ as $N \to \hat{N}$. Consequently E remains bounded and a limit cycle ensues (Figures 6.15 and 6.16).

It is necessary to point out that whenever N approaches zero our conclusions can only be suggestive of what actually happens. In formulating population models in Chapter Four, it was assumed that the population is large enough to permit a continuous model to be used. In the vicinity of zero, the validity of this assumption begins to break down, and so we must be careful in the interpretation of the results. This caution applies *mutatis mutandi* to all the population models in this book.

6.4 The Flywheel Governor

One of the most significant developments of the early industrial revolution was the invention of the rotary steam engine in the eighteenth century by James Watt. In order to keep the engine running at a constant angular speed in the face of varying loads, which would have a tendency to speed up or slow down the engine, a feedback control device was installed known as the centrifugal governor. It regulated the flow of steam to the engine, increasing the flow if the engine starts to slow and decreasing it otherwise.

The idea of this device is illustrated in Figure 6.17. The engine is represented by a flywheel that rotates at an angular speed ω. This rotation is dictated by the amount of steam that is available. Connected to the flywheel by a set of gears is a spindle that is made to rotate faster, at an angular speed $n\omega$. Connected to the spindle is a movable hinge

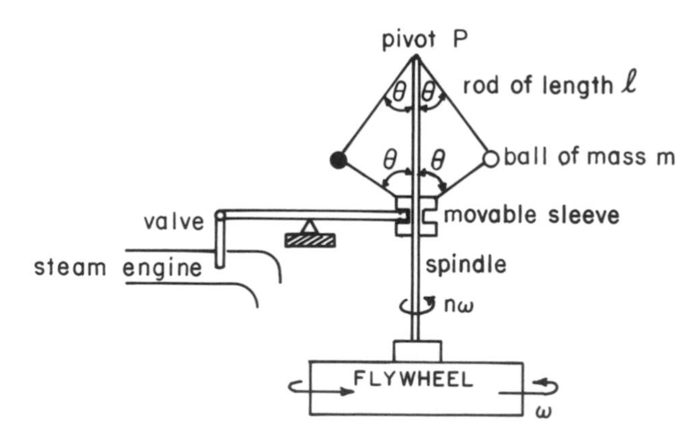

Figure 6.17 The centrifugal governor attached to a flywheel, as seen in a sideview.

attached to two identical rods of length l (of negligible mass) at the end of which are balls of mass m. As the rod rotates so does the rod–ball configuration. With a change in speed ω, the angle θ between the rods and the vertical position of the spindle also varies. In fact, the rod–ball set up is conceptually the same as the rod–bob configuration in the rotating pendulum discussed in Example 6.5. The only difference is that there are now two rods both linked to a movable sleeve that is fitted onto the spindle. As the rods rise with increasing ω so does the sleeve. But the sleeve is connected to a valve that closes to decrease the flow of steam as the sleeve rises. As the sleeve falls, the valve opens to admit more steam. In this way, the flywheel speed is regulated by decreasing steam in response to higher speeds and just the opposite in the contrary case. This is a feedback control system in the sense of Section 3.4.

It is desired to keep the flywheel rotating at a constant speed ω_0. A deviation of ω from ω_0 either raises or lowers the rods or the governor. This alters the angle θ, which raises or lowers the movable sleeve on the spindle, which, in turn, closes or opens the steam valve. As Figure 6.17 shows, the vertical movement of the sleeve is proportional to $\cos\theta$. Since the up and down movement regulates the flow of steam, θ is a control variable.

To derive the equations of motion, first note that the angle θ satisfies Equation 6.22 since the same force diagram applies to each of the rod–ball assemblies. Because the pivot at P supplies a frictional resistance, we can restate the equation as

$$\ddot{\theta} = (n\omega)^2 \cos\theta \sin\theta - \frac{g}{l}\sin\theta - r\dot{\theta}. \tag{6.32}$$

If I is the moment of inertia of the flywheel, then Newton's second law for rotational motion is

$$I\dot{\omega} = P_1 - P, \tag{6.33}$$

where P_1 is the torque due to the action of the steam and P is the torque due to the variable load on the flywheel. The torque P_1 is the sum of $\overline{P}_1$, at the desired rotational speed ω_0, plus an additional amount due to the deviation of ω from ω_0. The valve moves up and down to admit more or less steam by an amount $d(\cos\theta - \cos\theta_0)$, where d is the length of the link from each ball to the sleeve and θ_0 is the angle corresponding to a spindle speed of $n\omega_0$. The additional torque due to the steam is proportional to the movement of the valve, and so

$$P_1 = \overline{P}_1 + \alpha(\cos\theta - \cos\theta_0) \tag{6.34}$$

for some constant of proportionality $\alpha > 0$.

Hence 6.33 may be written as

$$I\dot{\omega} = \alpha\cos\theta - F, \tag{6.35}$$

where

$$F = (P + \alpha\cos\theta_0) - \overline{P}_1.$$

By letting $x_1 = \theta$, $x_2 = \dot{\theta}$, and $x_3 = \omega$, Equations 6.32 and 6.35 can be written as a third-order system:

$$\begin{aligned} \dot{x}_1 &= x_2 \\ \dot{x}_2 &= (nx_3)^2\cos x_1 \sin x_1 - \frac{g}{l}\sin x_1 - rx_2 \\ \dot{x}_3 &= \frac{\alpha\cos x_1}{I} - \frac{F}{I}. \end{aligned} \tag{6.36}$$

Rewrite the last equation as

$$\dot{x}_3 = k(\cos x_1 - \rho)$$

in which $k = \alpha/I$ and $\rho = F/\alpha$. Then the single equilibrium of 6.36 occurs at the point $\overline{\mathbf{x}}$ whose components in $\mathbf{R}^3$ are

$$\begin{aligned} \overline{x}_1 &= \arccos\rho \\ \overline{x}_2 &= 0 \\ \overline{x}_3 &= \frac{\beta}{\sqrt{\rho}} \end{aligned}$$

with $\beta = \dfrac{1}{n}\sqrt{\dfrac{g}{l}}$.

The linearized system corresponding to 6.36 has a Jacobian

$$A = \begin{pmatrix} 0 & 1 & 0 \\ (nx_3)^2(\cos^2 x_1 - \sin^2 x_1) - \dfrac{g}{l}\cos x_1 & -r & 2nx_3\cos x_1 \sin x_1 \\ -k\sin x_1 & 0 & 0 \end{pmatrix}.$$

The characteristic polynomial $\text{Det}(A - \lambda I) = 0$ can be written as the cubic equation

$$\lambda^3 + r\lambda^2 - \lambda\left[(nx_3)^2(\cos^2 x_1 - \sin^2 x_1) - \frac{g}{l}\cos x_1\right]$$
$$+2knx_3 \sin^2 x_1 \cos x_1 = 0. \tag{6.37}$$

Using the fact that

$$\sin^2 \bar{x}_1 = \sin^2(\arccos\rho) = 1 - \cos^2(\arccos\rho) = 1 - \rho^2,$$

the characteristic polynomial evaluated at $\bar{\mathbf{x}}$ gives the equation

$$\lambda^3 + r\lambda^2 + \frac{(n\beta^2)(1-\rho^2)\lambda}{\rho} + 2kn\beta\sqrt{\rho}\,(1-\rho^2) = 0. \tag{6.38}$$

It is not easy to compute the roots of this cubic explicitly in order to obtain the eigenvalues λ_i, $i = 1, 2, 3$, and so we proceed indirectly. In order to profit from Theorem 6.2, it is necessary that there be one real and two complex (conjugate) eigenvalues. The most general cubic with $\lambda_i = \sigma \pm iq$ $(i = 1, 2)$ and λ_3 real is $\prod_{i=1}^{3}(\lambda - \lambda_i)$, which gives us

$$\lambda^3 - (2\sigma + \lambda_3)\lambda^2 + (\sigma^2 + q^2 + 2\sigma\lambda_3)\lambda - \lambda_3(\sigma^2 + q^2) = 0. \tag{6.39}$$

In particular, Re λ_i should be zero where the bifurcation point occurs. At $\sigma = 0$, 6.39 becomes

$$\lambda^3 - \lambda_3\lambda^2 + q^2\lambda - \lambda_3 q^2 = 0. \tag{6.40}$$

Comparison of the coefficients of 6.40 and 6.38 reveals that

$$\lambda_3 = -r$$
$$q^2 = \frac{(n\beta)^2(1-\rho^2)}{\rho}$$
$$\lambda_3 q^2 = -2n\beta k\sqrt{\rho}\,(1-\rho^2).$$

From the last two relations, it follows that

$$-\frac{\lambda_3(n\beta)^2(1-\rho^2)}{\rho} = 2n\beta k\sqrt{\rho}\,(1-\rho^2),$$

and so, using the fact that $\lambda_3 = -r$, the specific value of r at $\text{Re}\,\lambda_i = 0$ is

$$\hat{r} = \frac{2k\rho^{3/2}}{n\beta}. \tag{6.41}$$

Let the bifurcation parameter μ be defined as $\mu = \hat{r} - r$. Then we expect a bifurcation to take place where $\mu = 0$ since that is where the two complex eigenvalues have zero real parts. It is easy to see that if $\mu \neq 0$, then the eigenvalues λ_1 and λ_2 are complex with nonzero real parts. In fact, $\text{Re}\,\lambda_i(\mu) < 0$ if and only if $\mu < 0$ (Exercise 6.5.10).

Since the eigenvalues vary with μ, let us denote them by $\lambda_i(\mu)$. Note that $\lambda_3(0) = \hat{r} < 0$, since $\mu = 0$ is equivalent to $r = \hat{r}$, and $\text{Re}\,\lambda_i(0) = 0$ for $i = 1, 2$. Also, $\text{Im}\,\lambda_i(0) \neq 0$ (Exercise 6.5.10). To apply Theorem 6.2, it suffices to look at $\text{Re}\,\lambda'(0)$, where prime indicates differentiation with respect to μ. Since $r = \hat{r} - \mu$, a differentiation of Equation 6.38 gives, for $i = 1, 2$,

$$\frac{d\lambda_i(\mu)}{d\mu} = \frac{1}{3 + \dfrac{2r}{\lambda_i} + \dfrac{(1-\rho^2)(n\beta)^2}{\rho\lambda_i^2}}.$$

When $\mu = 0$, we know that

$$\frac{(n\beta)^2(1-\rho^2)}{\rho} = q^2$$

and $\lambda_i = \pm iq$ ($\lambda_i^2 = -q^2$). Therefore,

$$\frac{d\lambda_i(0)}{d\mu} = \frac{\left(1 \pm \dfrac{i\hat{r}}{q}\right)}{2\left[1 + \left(\dfrac{\hat{r}}{q}\right)^2\right]}.$$

It follows that $\text{Re}\,\lambda_i'(0) > 0$. We therefore conclude that if $|\mu|$ is small enough, there exist limit cycles. It is known physically that when $\bar{\mathbf{x}}$ becomes unstable as μ increases past the bifurcation point, the governor begins to oscillate, a motion that has been called "hunting". This suggests that a stable limit cycle exists for $\mu > 0$.

In order to interpret the meaning of the instability condition $\mu > 0$, observe that this is equivalent, from 6.41, to the condition

$$r < \frac{2k\rho^{3/2}}{n\beta}. \tag{6.42}$$

Recall that r represents frictional resistance of the pivot divided by the mass m of the balls (as the derivation of 6.19 shows), and so an increase in m or a decrease in friction can contribute to decreasing r. Also, a lowering of flywheel inertia I will increase k. Any and all of these factors can lead to 6.42. Apparently what happened in the nineteenth century, as Watts original device became perfected, is that the improved machining of engine parts led to a reduction of friction while an increase in engine speeds meant a reduction in flywheel inertia. Also, as the engine size increased so did the mass of the balls used in the governor. As a result, flywheel governors became less reliable as time went on, with "hunting" becoming a more common problem. The breakdown of equilibrium stability and the existence of limit cycles seems to confirm the physically observable phenomena.

Since flywheel speed ω has a value proportional to $\rho^{-1/2}$ at equilibrium (where $\omega_0 = \bar{x}_3$) and because ρ depends directly on the load, it follows that ω decreases as the load increases. Moreover, $\theta_0 = \bar{x}_1 = \arccos \rho$ diminishes. Eventually, the load exceeds the ability of the steam engine to perform. With the valve fully open, the engine has slowed to halt. From 6.31 we see that this happens if ρ exceeds unity since it implies $\dot{\omega} < 0$. Therefore, a useful equilibrium requires that $\rho < 1$.

As θ varies so does ω, from Equations 6.35. However, $\ddot{\theta}$ depends on ω, as we see from 6.32. Hence, the variable θ is a feedback control since it changes as the state ω varies.

6.5 Exercises

6.5.1 Carry out a linearized stability analysis about the equilibrium of Equation 6.2 in which $g(\dot{x} - v)$ is given by Figure 6.3.

What happens if ω^2 is large enough?

6.5.2 Consider the following generalization of Equation 6.7

$$\phi'' + f(\phi) + c\phi' = 0$$

in which $f(\phi) \geq 0$ for $0 \leq \phi \leq 1$ and $f'(0) > 0$, $f'(1) < 0$, and $f(0) = f(1) = 0$. Using an argument similar to that in Example 6.3, show that there is a bounded invariant region Ω for the flow of this equation containing the equilibria $\begin{pmatrix} 1 \\ 0 \end{pmatrix}$ and $\begin{pmatrix} 0 \\ 0 \end{pmatrix}$.

6.5.3 Show that the system

$$\dot{x} = \mu x + y$$
$$\dot{y} = -x + \mu y$$

has a Hopf bifurcation at $\mu = 0$, but the only nontrivial cycles are those surrounding the origin at $\mu = 0$.

6.5.4 Show that the equation $\dot{z} = \mu z - z^3$ has a pitchfork bifurcation at $\mu = 0$. Describe the orbits.

6.5.5 Establish the global a.s. of the equilibrium in Equation 6.22, whenever $\omega < \sqrt{g/l}$.

6.5.6 The system consisting of Equations 6.26 and 6.27 has a globally a.s. equilibrium. Show this and interpret. Why would this result be called economic over-fishing if $c/p < K/2$?

6.5.7 Suppose $f(N)/N$ has a maximum at $N = \hat{N} > 0$. Show that $f''(\hat{N}) < 0$.

6.5.8 Recall Exercise 4.5.13. A somewhat artificial but not unreasonable model that includes satiation leads to the following equations:

$$\dot{x}_1 = \sigma - Qx_1 - \frac{x_1 x_2}{\alpha + x_1}$$

$$\dot{x}_2 = Qx_2 + \frac{x_1 x_2}{\alpha + x_1} - \frac{x_2 x_3}{\alpha + x_2}$$

$$\dot{x}_3 = -Qx_3 + \frac{x_2 x_3}{\alpha + x_2}$$

with σ, Q, and α all positive. This three-dimensional system can be reduced to two dimensions by defining a new variable $z = x_1 + x_2 + x_3$. Find and solve the differential equation satisfied by z and interpret this geometrically. Use this result to reduce the above system into an equivalent system involving only x_1 and x_2.

6.5.9 Find the equilibria of the system in the previous exercise in terms of x_1 and x_2. Show that if $Q \geq 1$, then there is only one equilibrium for all $\sigma > 0$ and it is locally a.s. For $Q > 1$, this equilibrium becomes unstable for σ large enough. When this happens, a new locally a.s. equilibrium appears. A painful computation shows that as σ increases further this equilibrium also loses stability and a third equilibrium manifests itself that is repelling for σ large enough. Deduce that a limit cycle exists in this case. Note that the equilibria are viable only if the population values x_i are nonnegative.
Warning: The computations are a bit tricky.
Interpret the significance of the above results.

6.5.10 Show, by comparison with 6.35, that the complex eigenvalues λ_1 and λ_2 of 6.34 depend on r in such a way that $\operatorname{Re} \lambda_i(r) \neq 0$ for $r \neq \hat{r}$ and $\operatorname{Im} \lambda_i(\hat{r}) \neq 0$. Also, of course, $\operatorname{Re} \lambda_i(\hat{r}) = 0$.

6.5.11 Recall the two prey–one predator model which was considered in Exercise 4.5.17. For simplicity, suppose that both victim species have growth rates and carrying capacities that are all identical to one, say. Choose the constants describing competitive advantage so that the nontrivial equilibrium is unstable in the absence of the predator P_3 (see the discussion in Section 4.3).

With P_3 present, however, the picture is different. If P_3 ravishes the victims P_1 and P_2 at different rates, then coexistence among all three species now becomes possible. Either a stable equilibrium exists or a limit cycle. A proof of this is difficult to achieve, but by computer experimentation some evidence of the truth of this statement may be gleaned. This is the first but not the last time in the present text that we make an appeal to the numerical computation of orbits. In Chapter Eight, it is seen as an important weapon if one is to achieve some understanding of otherwise intractible models.

The ecological significance of this coexistence is that predation can sometimes act to ensure the continuation of diverse species rather than being an agent of extermination.

In the special case in which P_3 preys at the same constant rate γ on both P_1 and P_2, show that even with P_3 present the nontrivial equilibrium continues to be unstable. This can be carried out explicitly.

6.5.12 The FitzHugh model that arises in neurophysiology is described by the equations

$$\begin{aligned} \dot{x} &= -f(x) - y + z \\ \dot{y} &= b(x - \gamma y) \end{aligned} \tag{6.43}$$

in which $f(x) = x(x - a)(x - 1)$ with $0 < a < 1$, $\gamma > 0$, and $b > 0$. The function $z(t)$ is zero for $t < 0$ and equal to a constant $q > 0$ for $t \geq 0$. It represents a stimulus, while x is the response. The response corresponds physically to a neuron pulse. For each q, there is an equilibrium solution to 6.43. Show that if

$$\gamma \leq \frac{3}{a^2 - a + 1},$$

then 6.43 possesses a unique rest point which is a.s. for all q less than some threshold value $\hat{q}$ and that a Hopf bifurcation takes place at $\hat{q}$. This model is suggestive of the fact that a large enough stimulus can trigger a repetitive firing of neuron pulses while below a certain threshold there is no response to the stimulus. In the next chapter, this threshold phenomenon is re-examined in a different way in terms of catastrophe theory.

Show that by a suitable change of variables 6.43 becomes qualitatively similar to the Van der Pol equations 2.29.

6.5.13 Let us modify the predator–prey model 4.21 by assuming that the predator has access to only those prey that find themselves on the periphery of some spatial region. That is, the predator acts solely along the edge of a given territory. The prey is assumed to be uniformly distributed in space.

Write down an appropriate model for the interaction between the two species. Show that a limit cycle exists about a nontrivial but unstable equilibrium whenever the prey carrying capacity is large enough. Interpret this result.

Chapter **Seven**

Bifurcation and Catastrophe

7.1 Fast and Slow

An important subclass of dynamical systems are the so called *gradient systems* which are described by the equations

$$\dot{\mathbf{x}} = -\nabla U(\mathbf{x}) \tag{7.1}$$

for $\mathbf{x}$ in $\mathbf{R}^k$. The expression ∇U is the gradient of a smooth scalar function U that is twice continuously differentiable, namely the vector in $\mathbf{R}^k$ whose components are $\partial U / \partial x_i$, $i \leq k$.

Let $\bar{\mathbf{x}}$ be the only equilibrium of 7.1 on some open set Ω. Along an orbit of this equation we have, by the chain rule of differentiation,

$$\dot{U}(\mathbf{x}(t)) = \nabla U \cdot \dot{\mathbf{x}}(t) = -\|\nabla U\|^2, \tag{7.2}$$

which is negative on Ω except at $\bar{\mathbf{x}}$ where it is zero. The sets in $\mathbf{R}^k$ defined by $U(\mathbf{x}) = \alpha$, for any real α, are level surfaces of U, and from multivariate calculus it is known that all nonzero gradient vectors $-\nabla U$ are orthogonal to these surfaces and point "downhill" in a direction of decreasing U. An immediate consequence of this is that there can be no

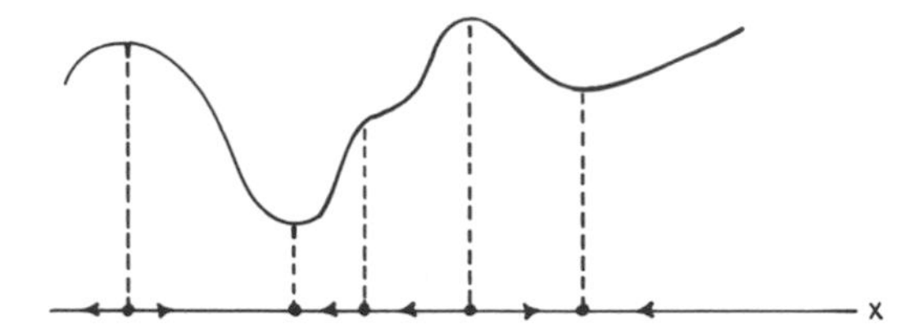

Figure 7.1 The gradient system 7.1 envisaged as a landscape in $\mathbf{R}^1$, in which the valleys are basins of attraction for the minima of a function U. The direction of orbits away from or towards the equilibria is indicated by arrows.

closed orbits. Observe, also, that if $\bar{\mathbf{x}}$ minimizes U on Ω, then ∇U is zero at $\bar{\mathbf{x}}$ and therefore the function $V(\mathbf{x}) = U(\mathbf{x}) - U(\bar{\mathbf{x}})$ is Liapunov on Ω. Even though we proved the Liapunov Theorem on $\mathbf{R}^2$ (Theorem 3.1), virtually the same argument goes through in $\mathbf{R}^k$, for any k. Hence, $\bar{\mathbf{x}}$ is a.s. and it attracts all orbits beginning in Ω. For similar reasons, a maximum of U is a repeller. Now the positive limiting set of any orbit of 7.1 consists of equilibria (consult the references to Chapter Seven for a proof of this assertion). Consequently $\mathbf{R}^k$ is divided up into basins of attraction for stable rest points, separated by the stable manifolds of saddle points. Figure 7.1 shows this for $k = 1$.

It will be convenient to refer to U as a potential function, even though we are outside of the context of the second-order dynamical systems treated in Chapter Two. The similarity in behavior between the function U considered there and those being discussed at present is evident from Example 3.4. Therefore, 7.1 will be considered as a model for dynamical systems whose goal, loosely speaking, is to minimize some potential function.

Consider now gradient systems which depend on a single parameter α or on two parameters α and β:

$$\dot{x} = -U'(x, \alpha)$$

or (7.3)

$$\dot{x} = -U'(x, \alpha, \beta),$$

where prime denotes differentiation with respect to x in $\mathbf{R}^1$. In physical applications, we think of the parameter(s) as changing slowly while x moves rapidly to its equilibrium. In effect, for each value of α (or α and β) we expect to find the system at or near a zero of U', which we denote by $\bar{x}(\alpha)$ or $\bar{x}(\alpha, \beta)$ to show the explicit dependence. Usually a small change in parameter induces a correspondingly small change in the rest state. However, for certain values of α, β the equilibrium can make a sudden and rapid jump in value. This fast–slow interaction between $\bar{x}$ and the parameters will be elaborated on below after giving two examples.

Example 7.1 Let

$$U(x, \alpha) = \frac{x^3}{3} - \alpha x \tag{7.4}$$

with x scalar valued. This function is exhibited in Figure 7.2 together with various cross sections in Figure 7.3. The set of (x, α) for which $\partial U/\partial x = 0$ is the parabola defined by $x^2 - \alpha = 0$. There are no equilibria for $\alpha < 0$. However, as α passes zero there is a pitchfork bifurcation of the type studied in Section 6.2 in which an unstable rest point at $x = 0$ (for $\alpha = 0$) gives rise to two new equilibria for $\alpha > 0$, one stable and the other unstable. In fact, this example was already encountered in Exercise 6.5.3 but in a slightly different setting.

The normal to the implicitly defined curve

$$g(x, \alpha) = \frac{\partial U}{\partial x} = x^2 - \alpha = 0$$

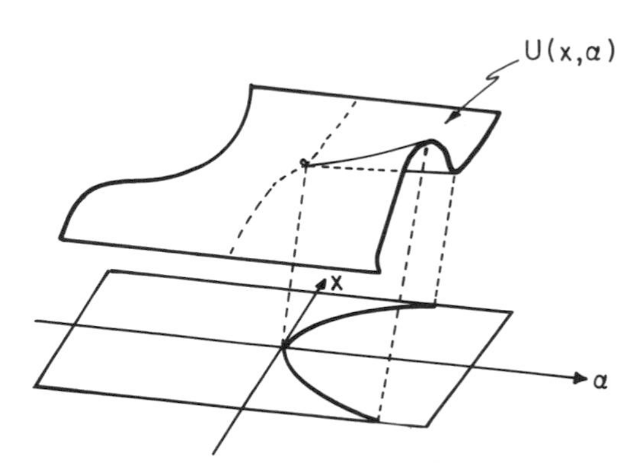

Figure 7.2 The function $U(x, \alpha) = x^3/3 - \alpha x$, shown elevated above the x, α plane. Actually, it cuts through the plane.

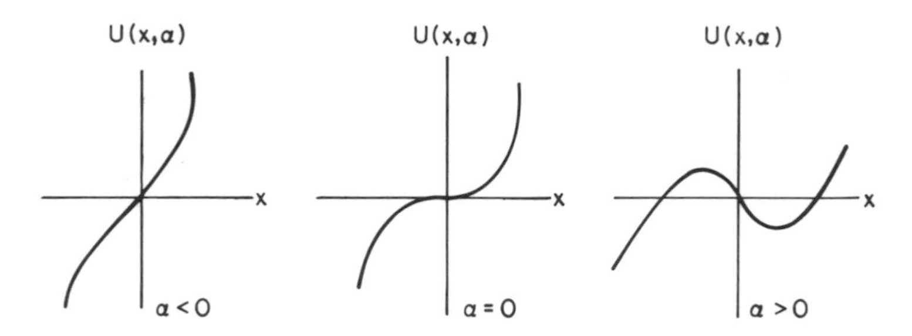

Figure 7.3 Cross-sections of the function U in the previous figure for different values of α. Note the inflection point at $x = \alpha = 0$. Here $U'' = 0$.

is the gradient vector

$$\begin{pmatrix} \dfrac{\partial g}{\partial x} \\ \dfrac{\partial g}{\partial \alpha} \end{pmatrix} = \begin{pmatrix} 2x \\ -1 \end{pmatrix},$$

which lies along the horizontal α axis when $x = 0$. This forces α to be zero at this point. In other words, the tangent vector to the curve is perpendicular to the α axis at $\alpha = 0$, precisely where the bifurcation takes place. Since $x(\alpha)$ changes its character suddenly, by folding outward to form a pitchfork, the bifurcation point is called a *fold catastrophe*. Observe that $\partial g/\partial x = \partial^2 U/\partial x^2$, and so the catastrophe takes place where the second derivative of U vanishes.

***Example** 7.2* This example is more interesting than the previous one and, as will be seen, serves as a prototype for all the models to be discussed later in this chapter. Let

$$U(x, \alpha, \beta) = \frac{x^4}{4} - \frac{\alpha x^2}{2} - \beta x. \tag{7.5}$$

with x again scalar valued. Consider the surface $\mathcal{M}$ in $\mathbf{R}^3$ defined implicitly by

$$\frac{\partial U}{\partial x} = g(x, \alpha, \beta) = x^3 - \alpha x - \beta = 0. \tag{7.6}$$

It is noteworthy that the equilibria $\bar{x}(\alpha, \beta)$ of

$$\dot{x} = -(x^3 - \alpha x - \beta) \tag{7.7}$$

satisfy 7.6 and hence lie on $\mathscr{M}$.

Small changes in α and β induce slight alterations in $\bar{x}(\alpha, \beta)$ except where the tangent to $\mathscr{M}$ is perpendicular to the α, β plane $\mathscr{N}$ or, equivalently, where the normal to $\mathscr{M}$ lies parallel to $\mathscr{N}$. The normal is given by the gradient vector

$$\begin{pmatrix} \dfrac{\partial g}{\partial \alpha} \\ \dfrac{\partial g}{\partial \beta} \\ \dfrac{\partial g}{\partial x} \end{pmatrix} = \begin{pmatrix} -x \\ -1 \\ 3x^2 - \alpha \end{pmatrix},$$

which has a zero component in the x direction precisely where

$$\frac{\partial g}{\partial x} = \frac{\partial^2 U}{\partial x^2} = 3x^2 - \alpha = 0.$$

Hence, sudden changes in equilibria occur on the set in $\mathbf{R}^3$ defined implicitly by $\partial U/\partial x = \partial^2 U/\partial x^2 = 0$, namely

$$\begin{aligned} x^3 - \alpha x - \beta &= 0 \\ 3x^2 - \alpha \qquad &= 0. \end{aligned} \tag{7.8}$$

Eliminating x from these equations gives

$$\beta^2 = \frac{4}{27}\alpha^3, \tag{7.9}$$

which represents a cusp-like curve C in the plane $\mathscr{N}$, called a *cusp catastrophe*. The surface $\mathscr{M}$ and curve C are represented in Figure 7.4. It is sometimes useful to reorient this picture, tantamount to taking the sign of β in 7.5 and 7.6 to be positive. This is also illustrated in Figure 7.4.

When α and β move across C, the corresponding $\bar{x}(\alpha, \beta)$ jumps from a lower (or upper) sheet of $\mathscr{M}$ upward (downward) to a different sheet. The projection onto $\mathscr{N}$ of the pleat of $\mathscr{M}$ is the interior of C. Outside of

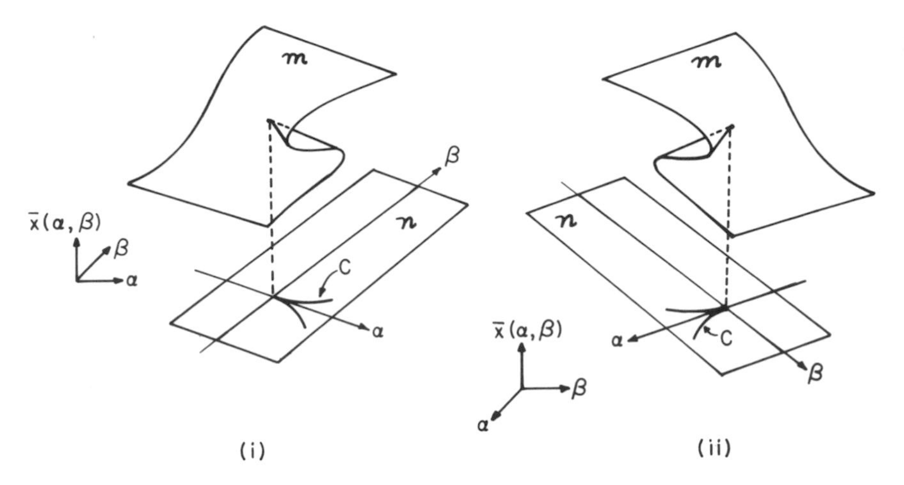

Figure 7.4 The surface $\mathcal{M}$ is defined implicitly by $x^3 - \alpha x - \beta = 0$ in (i). The points of $\mathcal{M}$ consist of equilibria of Equation 7.7. $\mathcal{M}$ is shown elevated above the α, β plane $\mathcal{N}$. Actually it cuts through this plane. The curve C defined implicitly by 7.9 is the projection of the edges of the pleat of $\mathcal{M}$ onto $\mathcal{N}$. Also shown in (ii) is an alternate representation when the surface is implicitly defined by $x^3/3 - \alpha x + \beta = 0$. In either case $\mathcal{M}$ is a triple-sheeted covering of $\mathcal{N}$ within C and single-valued outside of C.

C, $\bar{x}$ is single valued. Within C, however, $\mathcal{M}$ is a triple covering of $\mathcal{N}$, and there are three equilibria. These occur as bifurcations as α, β moves inwardly across C while an outward move coalesces the equilibria into one. Figure 7.5iii shows what happens as β varies with α fixed at some positive value. The picture displays a cross section of the surface $\mathcal{M}$ defined by 7.7. Between β_1 and β_2 there are three equilibria, but the

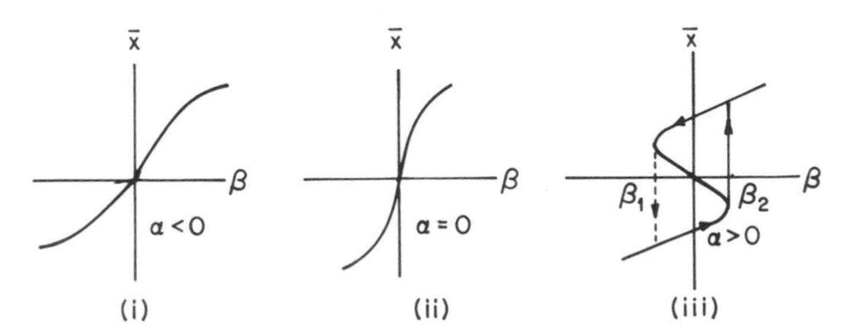

Figure 7.5 Cross-sections of the surface $\mathcal{M}$ defined by $x^3 - \alpha x - \beta = 0$, for different α. In (iii) the arrows indicate the $\bar{x}$ moves slowly upward as β increases from β_1 to β_2, where it then jumps rapidly up. The opposite effect takes place at β_1 as β decreases from β_2.

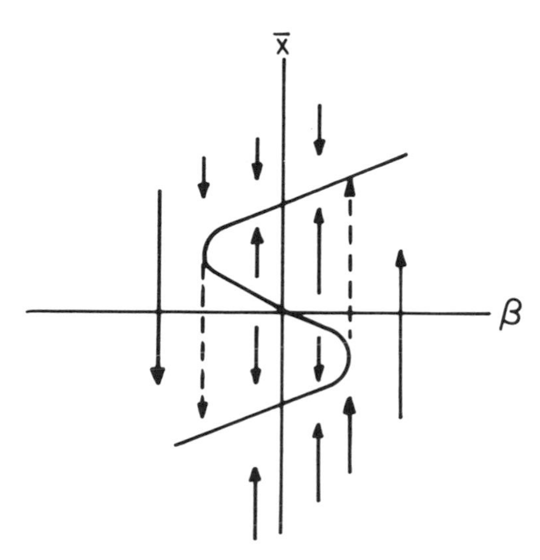

Figure 7.6 Direction of orbits toward their equilibria in case (iii) of Figure 7.5. The middle branch is evidently unstable.

middle one is never attained by any orbit as β increases. Initially, $\bar{x}$ is on the lower branch of $\mathcal{M}$, but when $\beta = \beta_2$, $\bar{x}$ jumps up as shown by the solid line. As β decreases $\bar{x}$ keeps on the upper branch of $\mathcal{M}$ until β reaches β_1, where it then jumps down (dotted line). The return path does not retrace the forward one, a phenomenon known as *hysteresis*.

Since $\beta = x^3 - \alpha x$ on the "S" shaped curve γ on Figure 7.5, it follows that if $\beta > \gamma$ ($\beta < \gamma$), then $\dot{x}$ is positive (negative). The orbits of 7.7 therefore move as indicated in Figure 7.6 in which it is apparent that the equilibria on the middle branch are unstable, while the other two are a.s.

Let us pause to note that if α is varied from negative to positive values, while β is fixed at zero, then we encounter a fold catastrophe at $\alpha = 0$. The bifurcation behavior of Example 7.1 is therefore a special instance of the present case.

These examples show that a catastrophe occurs either where new equilibria appear by bifurcation or existing equilibria coalesce into one. For two parameters, the scenario is this. Think of the orbit as varying in fast time according to 7.3 until it reaches an equilibrium $\bar{x}(\alpha, \beta)$. Thereafter, α and β vary in slow time causing $\bar{x}$ to either change smoothly or to jump quite suddenly to a new value. This jump occurs

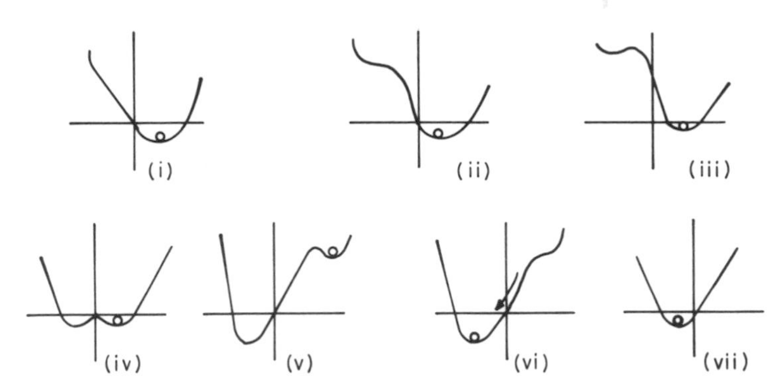

Figure 7.7 The shifting landscape of some arbitrary potential function U on $\mathbf{R}^1$ as β varies and α is some fixed value. There is a single equilibrium, a stable minimum, in (i). A bifurcation takes place in (ii) as an inflection point is created. This gives rise to three equilibria in (iii)–(v). However, the orbit remains at or near the original equilibrium, even though it moves slightly as β varies, until (vi) at which juncture all three points coalesce. In (vii) there is again an isolated stable equilibrium. Moving these images in reverse has the orbit jumping back to the original equilibrium only after we have backtracked to (ii). This illustrates the hysteresis effect.

whenever the second derivative U'' vanishes at some inflection point of U. Geometrically, this corresponds to the tangent becoming perpendicular to the α, β plane $\mathcal{N}$.

The same kind of discontinuous behavior can be observed for potentials other than 7.5, as shown in a series of "snapshots" of some arbitrarily chosen U (Figure 7.7). This illustrates what is truly the most remarkable part of the story, namely that the behavior of virtually any smooth potential U at the critical points where $U' = 0$ is not substantially different than that of the particular function 7.5. In other words, the cusp catastrophe of the surface $\mathcal{M}$ already exhausts all the possibilities. We want to make this statement more precise by stating a rather deep theorem of R. Thom. To do so requires that some technical notions be introduced first. The several applications given later in this chapter will clarify the significance of the theorem.

By a smooth function, we now mean one that is infinitely differentiable. Let $\mathcal{F}$ be the class of potential functions U which depend smoothly on x in $\mathbf{R}^1$ and (α, β) in $\mathbf{R}^2$. When we say that a property is generic in $\mathcal{F}$ what is meant, in an informal way, is that it holds with rare exception for all U in $\mathcal{F}$. Technically, a topology is introduced in $\mathcal{F}$, and a generic property is then one which holds on an open and dense subset (recall, in this connection, the discussion in Section 3.3). Two subsets A and B of $\mathbf{R}^2$ are said to be smoothly deformable to each other, indicated by $A \leftrightarrow B$, if there is a smooth and invertible function taking one into the other.

In what follows, the surface $\mathcal{M}$ and the curve C are those defined by 7.6 and 7.9 and shown in Figure 7.4. The following theorem of R. Thom is now stated without proof.

Theorem 7.1 *The following properties are generic on* $\mathcal{F}$:

(*i*) *For* U *in* $\mathcal{F}$ *the surface* $\hat{\mathcal{M}}$ *in* $\mathbf{R}^3$ *defined implicitly by* $\dfrac{\partial U(x, \alpha, \beta)}{\partial x} = 0$ *is smooth.*

(*ii*) *Let* P *be the orthogonal projection of* $\hat{\mathcal{M}}$ *onto the* α, β *plane, and let* $\hat{C}$ *be the image under* P *of the points where the tangent to* $\hat{\mathcal{M}}$ *is perpendicular to* $\mathcal{N}$. $\hat{C}$ *is called the* catastrophe set. *One can then smoothly deform* $\hat{\mathcal{M}}$ *locally into* $\mathcal{M}$ *in such a way that* $\hat{C}$ *is also smoothly deformed into the cusp-like curve* C. *The following diagram therefore holds on a suitable neighborhood of every point of* $\hat{\mathcal{M}}$:

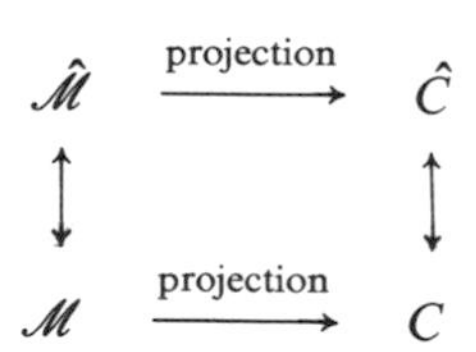

(*iii*) *Let* U *be perturbed slightly to some* $\tilde{U}$. *Then the surface* $\hat{\mathcal{M}}$ *can be smoothly deformed locally to the corresponding surface* $\tilde{\mathcal{M}}$ *of* $\tilde{U}$ *in such a way that the catastrophe sets* $\hat{C}$ *and* $\tilde{C}$ *are also smoothly deformable into each other. The following diagram therefore holds on a suitable neighborhood of every point of* $\hat{\mathcal{M}}$:

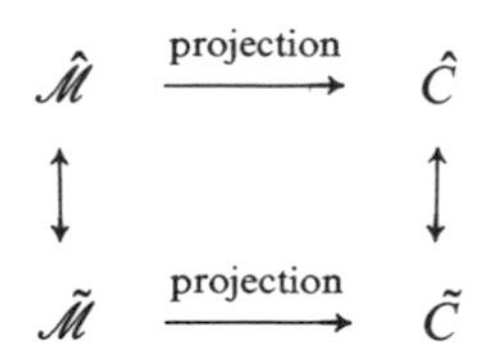

Stated roughly, the surface $\hat{\mathcal{M}}$ of a typical potential U is smoothly deformable into the standard surface $\mathcal{M}$, and its catastrophe set $\hat{C}$ is smoothly deformable into the cusp C. Moreover, $\hat{\mathcal{M}}$ is robust with respect to small changes in U.

We have given a somewhat weak statement of Thom's theorem. There is a more general version that applies to potentials that depend on as many as five parameters and to $\mathbf{x}$ in $\mathbf{R}^k$, for any k. The exact statement does not concern us here except to note that it classifies all possible catastrophes as belonging to a finite set of standard cases. For more than two parameters, there are several more complicated forms that a surface of the type $\partial U/\partial x = 0$ can take on, and the classification is not quite so simple as in the situation treated by us. Incidentally, for a single parameter the theorem says that the fold catastrophe is the typical occurrence.

It is useful to point out that the above theorem does not require that one know U explicitly. It applies generically to all smooth potentials. Moreover, U need not be tied to a specific differential equation as in 7.3. As long as one believes that the internal dynamics of a system is such that it tends to minimize some (generally unknown) smooth function U that depends on two parameters, then a cusp catastrophe may be expected. In the examples given later in the chapter, it is indeed true that the dynamics are governed by a gradient-like system but our point is that the validity of the theorem is really independent of knowing the specific manner in which U is minimized. The important modeling lesson here is that if we are willing to accept the existence of a potential that varies with two slowly changing parameters, then all motion is toward rest points that lie on a cusp-like surface.

The distinguishing feature of Figure 7.4 is that a stable equilibrium will find itself in essentially one of two states depending on whether it is on the upper or lower sheet of the pleated surface. It is therefore *bistable*. Moreover, a movement from one sheet to the other does not retrace itself as the parameters reverse their direction. This is the phenomenon of *hysteresis*, as noted earlier.

In some modeling studies, the issue is less whether a given equilibrium is stable or not, but rather if the system is resilient enough to persist in the face of inevitable change in its environment. In systems with gradient dynamics, a stable equilibrium $\bar{x}$ will alter with the slow variables α and β, which represent gradual environmental shifts. The basin of attraction of $\bar{x}$ may also vary. As long as an orbit continues to hover about $\bar{x}$, change is perceived to be mild and orderly: a slight shift in α and β gives a slight movement in $\bar{x}$. However, if $\bar{x}$ eventually gets very close to the boundary of its basin of attraction, then a further mild perturbation in α and β can catapult the orbit into an entirely different basin of attraction. This is particularly significant in the community of living things where arguably the success of a given species is its ability to persist even while its habitat is in a state of flux. The unknown environment, represented by α and β, either fluctuates at random or it moves inexorably so as to alter the conditions of survival. If it moves at random, for example, then a severe enough fluctuation can cause a sudden shift into a new and unfavorable environmental basin where survival is difficult or even impossible. It has been conjectured that a drastic change in climatic conditions at the end of the Cretaceous period may have produced a habitat hostile to the continuation of the dinosaurs. They became extinct. Is this an example of cusp catastrophe?

We have discussed a single equation in x for the dynamics of our gradient system in which the parameters α and β are not necessarily under our control. However, there could be a feedback linkage whereby these values actually depend on the current state x. In the examples given in the next few sections, this is actually the case for either α or β (or both), and the linkage is accomplished by introducing some additional differential equations. It works like this: α and β determine the orbit x. As x varies, it in turn influences the values of α and β via the additional differential equation(s). Since orbits are assumed to move quickly to their rest points, these relationships prescribe the flow on $\mathcal{M}$, including catastropic jumps, if any. If the dynamics of α and β are chosen properly, one can end up with cyclic motion in which x wanders for a time on one sheet of $\mathcal{M}$, quickly moves to another sheet for another sojourn, and then just as quickly returns to the original sheet where the process is repeated. Since the orbit lingers between jumps at one of essentially two values (it is bistable), it is possible to study the overall dynamics by first solving for x with α and β effectively constant. This determines the bistable values of the state variable. Then fix x at one of these levels and allow α or β (or both) to vary. By this ploy, the complications of the linked system are made more palatable.

Before closing this section, it may be useful to contrast the bifurcations that take place in gradient systems to the Hopf bifurcation studied in the previous chapter. There, as a single parameter changes, an equilibrium loses its stability and gives way to a closed orbit. In the present setting, however, a stable rest point gives rise instead to other rest points as a parameter passes some critical value.

7.2 The Pumping Heart

The human heart is a pump that takes in blood from the lungs, where it has been re-oxygenated, and then sends it back out to the rest of the body. We are not interested in following the exact details of how it functions but only in its overall dynamical behavior as a pump, and so our description is but a caricature of its actual physiology.

As a gross simplification, the heart consists of a chamber with inlet and outlet valves. Initially, the pump is in its relaxed state, called the diastole. Then an electrochemical stimulus moves over the heart causing its muscle fibres to contract, thereby pushing the blood out. This happens slowly at first in order to ensure no damaging backflow. At some high enough level of stimulus, however, the fibres contract suddenly for a final big push of blood through the body. The contracted state is called systole. At this juncture, the external stimulus temporarily ceases and the chamber begins to fill again, slowly at first as the fibres begin to relax, and then quite rapidly back into diastole. As this process repeats itself, the length x of muscle fibre moves through a cycle (Figure 7.8).

As it happens, there is a mathematical model that imitates Figure 7.8. Consider the Van der Pol equations

$$\dot{x} = v - \mu\left(\frac{x^3}{3} - x\right) \qquad (7.10)$$

$$\dot{v} = -x.$$

In Section 6.2, it was seen that a unique limit cycle exists for all $\mu > 0$.

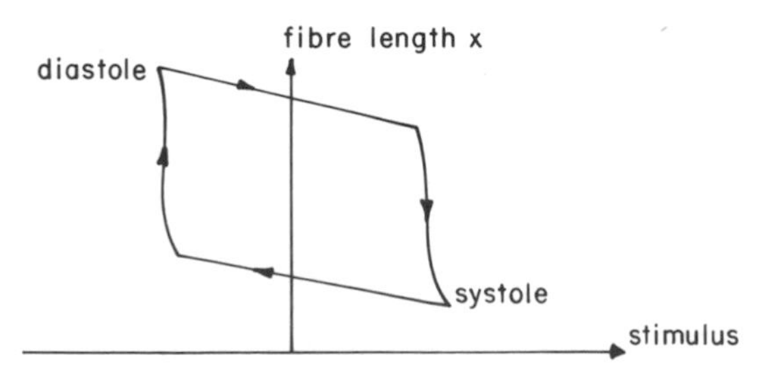

Figure 7.8 Schematic of the pumping action of the heart.

What does it look like? For small μ, the equations approximate those of the harmonic oscillator $\ddot{x} + x = 0$, whose orbits are circles about the origin, but for μ large the situation is quite different. Let $u = v/\mu$ to obtain the equivalent system

$$\dot{x} = \mu\left[u - \left(\frac{x^3}{3} - x\right)\right]$$
$$\dot{u} = -\frac{x}{\mu}. \tag{7.11}$$

Denote by γ the curve $u = x^3/3 - x$. Then $\dot{x} > 0$ for all u above γ, and $\dot{x} < 0$ for u below. Since μ is assumed to be large, $\dot{u}$ is small in magnitude while $\dot{x}$ is large for all u away from γ. This means that the orbit moves very rapidly along paths on which u is nearly constant until it approaches γ. Then $\dot{x}$ begins to decrease in magnitude and the orbit slows down. After this, it follows γ until it reaches the points A or B shown in Figure 7.9, where it jumps off and a cycle is repeated. Motion along γ is of course relatively slow. The direction of the orbit is determined by the fact that $\dot{u}$ is negative for x positive, and vice-versa. The part of γ between A and B is clearly repelling, and the orbit never actually moves along this portion.

There is a resemblance between the two preceding figures. In fact, if Figure 7.9 is rotated counterclockwise so that x is put on the vertical axis and if the picture is then moved upward, the Van der Pol cycle appears to model the heartbeat. It this it? Not quite, although we are getting close. To begin with, let us identify the electrochemical stimulus with a parameter β and let $u = -\beta$ to get the proper orientation shown

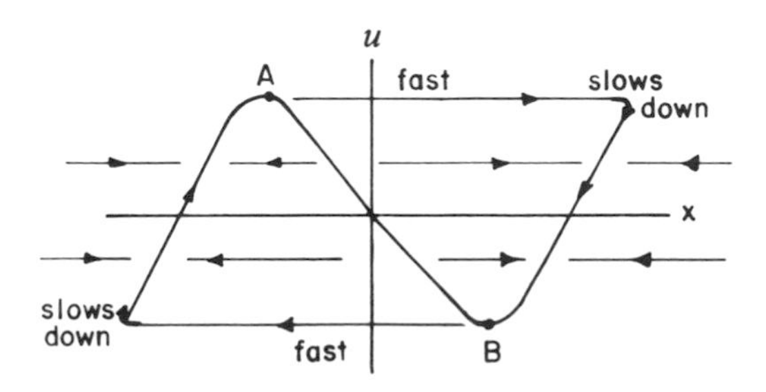

Figure 7.9 Orbit portrait for Equations 7.11 when μ is very large.

in Figure 7.8. This gives us the equations

$$\dot{x} = \mu\left(-\beta - \frac{x^3}{3} + x\right)$$
$$\dot{\beta} = \frac{x}{\mu}. \tag{7.12}$$

Next, we take into account the experimental fact that if the heart muscle fibre is already slightly stretched before beating begins, then a larger beat will result. The stretching is caused by tension which itself results from increased blood pressure in moments of stress. When this happens, the blood vessels in the cardiovascular system contract and push blood into the heart chamber at a faster rate. This causes the fibres to stretch, and the heart contracts more forcefully than usual. On the other hand, if tension is low, the beat will be weaker.

It was proposed by Zeeman that the model be modified by incorporating tension as an additional parameter α. Motivated by the cusp catastrophe surface, whose cross-sections look like Figure 7.8, the proposed model for fibre length x now becomes

$$\dot{x} = \mu\left[-\beta - \left(\frac{x^3}{3} - \alpha x\right)\right]. \tag{7.13}$$

This equation, which is essentially the same as 7.7, reproduces the cusp surface shown in Figure 7.4ii in which the pleated portion corresponds to the bistable states where the heart muscle is either relaxed or contracted. Movement between these states is rapid since μ is large. This is shown in Figure 7.10. In interpreting this figure, it is of course understood that actual fibre length, being positive, is obtained from x by adding some positive constant. In effect, one must imagine the entire surface translated upward. Also, no special significance is attached to the fact that α and β can be negative. They simply represent parameters that vary along some range of values that we conveniently locate in the $\mathcal{N}$ plane.

Let us see if the model corresponds to the known facts of the heart pump. Suppose tension α is positive. As stimulus β increases the muscle contracts, slowly at first while x is on the upper sheet, and then quite suddenly as x jumps to the lower sheet. When tension is low the jump is small, representing a small beat. As tension increases, the beat also gets larger. Now the stimulus takes on only a limited range of values, with β_1

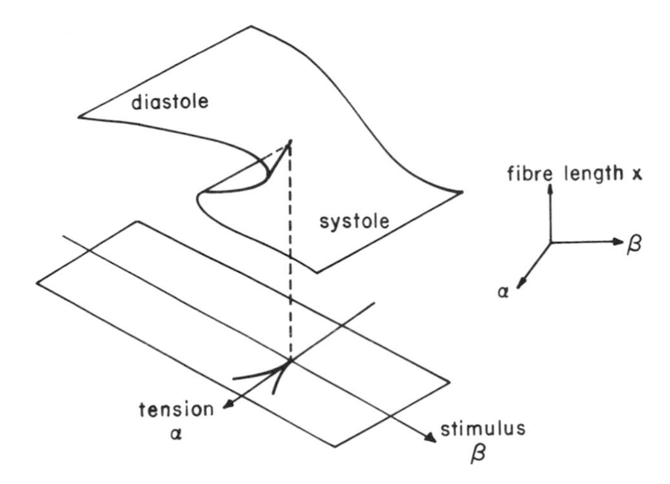

Figure 7.10 The cusp catastrophe model of the heart beat according to Equation 7.13. The surface actually passes through the α, β plane, but it is shown here as above the plane.

at diastole and a maximum β_2 at systole. It follows that under conditions of high tension the pleated part of the catastrophe surface will be so pronounced that as β moves from β_1 to β_2 the corresponding movement of x will not take it far enough to jump off the now more elongated upper sheet. Consequently, the fibre remains in close to diastole and moves very little, resulting in a feeble heart beat. In effect, this represents cardiac failure. This can indeed occur to a person with high blood pressure who receives a sudden shock thereby overstretching the muscle fibres. This would be indicated by a high value of α. On the other hand, if α is quite low (negative in our model), then x heaves sluggishly on that part of the cusp surface where there is no pleat. This occurs, for example, when the bloodstream is made to bypass the heart during an operation so that tension drops. The cusp model does, therefore, plausibly reflect the known behavior of the heartbeat under different levels of tension. These observations are summarized in Figure 7.11 by taking cross sections of Figure 7.10.

In order to reproduce the cyclic pumping action shown in Figure 7.11, we need to demonstrate how x depends on the stimulus. The stimulus is caused by some external agent whose underlying biochemistry need not be explicitly identified except to note that it induces β to change from a value of β_1, where the heart is in a state of rest (diastole), to a maximum

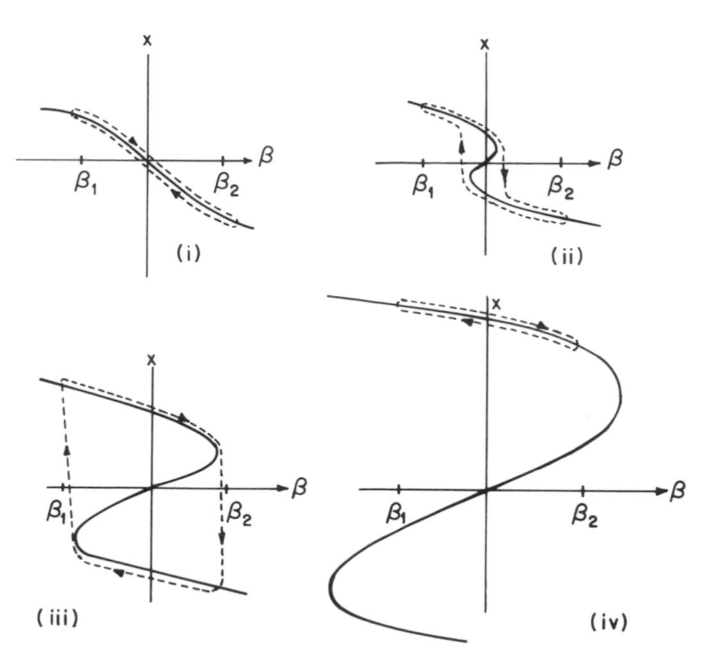

Figure 7.11 Cross-sections of the cusp model in Figure 7.10 for different values of the tension α. The surface is now correctly pictured as passing through the α, β plane. In (i) tension is very low, with α negative, while (ii) shows low tension but with α now positive but small. High tension occurs in (iii) for a larger α, while the very high tension in (iv) corresponds to an even larger α.

value β_2, which corresponds to systole. The second of Equations 7.12 shows that $\dot{\beta} = x/\mu$. Therefore, β increases while x is on the upper branch of the cusp surface, since x is positive there, and it decreases on the lower. This generates a closed orbit which is a cyclic solution of the modified Van der Pol equations discussed in Example 6.4:

$$\begin{aligned} \dot{x} &= \mu\left[-\beta - \left(\frac{x^3}{3} - \alpha x\right)\right] \\ \dot{\beta} &= \frac{x}{\mu}. \end{aligned} \tag{7.14}$$

These equations have an unique equilibrium at the origin, which is inaccessible to an orbit which begins, say, at systole since this rest point is unstable. However, this is a flaw in our model because we know that diastole is actually a stable equilibrium state of the heart toward which the heart muscle, after completing its contraction at systole, would eventually return. In short, the model should have an attracting equilibrium at $\beta = \beta_1$, where fibre length has some value $\hat{x}$, rather than an unstable rest point at the origin. To mimic this behavior, let us change the equation for β to read

$$\dot{\beta} = \frac{x - \hat{x}}{\mu}. \tag{7.15}$$

This provides the appropriate feedback linkage between x and β in which β is defined to have the value β_1 when x is at $\hat{x}$. It is easy to check that $\begin{pmatrix} \hat{x} \\ \beta_1 \end{pmatrix}$ is a.s.

The external electrochemical stimulus forces β to increase to β_2. This obliges x to be dislodged from its stable rest position. As a result x decreases and, depending on the value of α, the fibres slowly contract until x jumps onto the lower sheet of the cusp surface, following the dynamics of 7.13. It continues to move on this sheet until β equals β_2, at which point the fibres have attained maximum contraction. The external stimulus is now temporarily removed and the system relaxes. β is now free to vary according to 7.15. This means that as long as x remains on the lower sheet β decreases. Eventually, x must jump to the upper branch to a point that is either above or below $\hat{x}$. For high tension, the jump is in fact expected to take x to a value above $\hat{x}$. Since $\hat{x}$ is stable, the orbit returns to this point and β goes to β_1. Then stimulus is then applied again and the cycle repeats itself. Thus, the feedback equation 7.15 dictates the return path to diastole while 7.13 establishes the movement of x toward systole. The dynamics of α is unknown and depends on factors external to the model itself. That is why no equation linking α to x is provided.

A portrait of the orbits of 7.13 and 7.15 in the absence of any stimulus is shown in Figure 7.12. This is constructed in much the same way as the dynamics of the fast Van der Pol equation was determined in the previous section, the main difference being the location of the equilibrium point. Each orbit cuts the curve γ defined by $\beta = \alpha x^3 - x/3$ horizontally since $\dot{x} = 0$ on γ. This implies a slight overshoot of the orbit into the region where $\beta > \gamma$ (and $\dot{x} < 0$). The path then hugs γ, moving slowly toward $\bar{x}$.

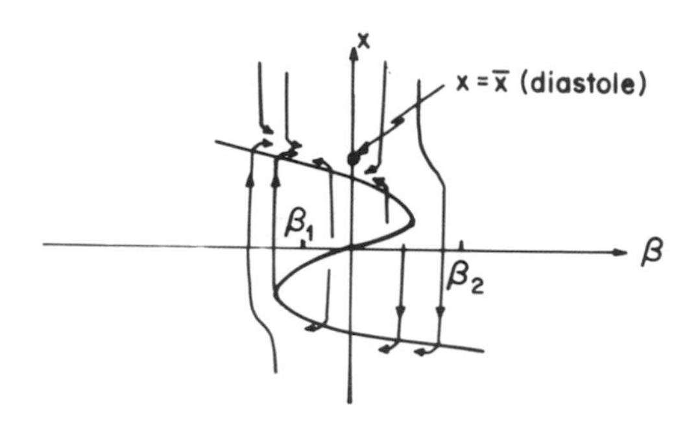

Figure 7.12 Orbit portrait of the heart model showing the relation between muscle fibre length x and stimulus level β for some sufficiently large positive tension α. The position $\beta = \beta_1$, $x = \bar{x}$ is an a.s. equilibrium.

When the stimulus is applied, this forces x to move away from $\hat{x}$, and the orbit is now altered to conform to the fact that x decreases as long as β increases. This is seen in Figure 7.13. Since x always moves quickly towards γ, the combined portraits of this and the preceding figure show that effectively a limit cycle is generated. This results in a rhythmic thumping of the beating heart.

In preparation for the next two sections, note that since the "S" shaped curve γ in Figures 7.12 and 7.13 is given by

$$\beta = \alpha x - \frac{x^3}{3}, \tag{7.16}$$

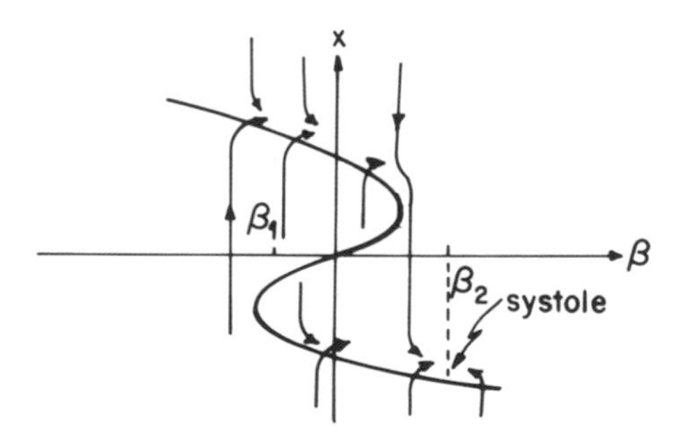

Figure 7.13 Orbit directions in a modified version of Figure 7.12 in which β is now forced to vary from β_1 to some β_2. The orbits follow a path toward a contracted state, the systole, which now becomes an attractor. After β reaches β_2, the system relaxes back into the mode shown in the previous figure in which diastole once again becomes the attracting state while systole becomes repelling. With x back at $\bar{x}$ the cycle repeats itself.

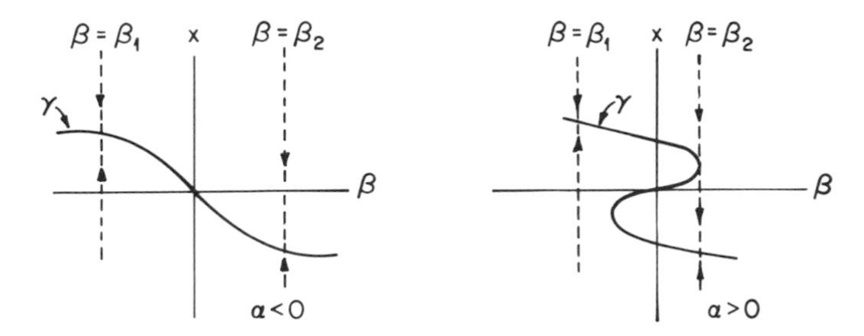

Figure 7.14 The surface $\beta = -x^3/3 + \alpha x$ for different β, with α positive or negative.

then for any given value of β, the Equation 7.16 has either one or three roots, representing stable or unstable equilibria. This is shown in Figure 7.14. For $\alpha < 0$, there is of course only one root for all β. Whenever the line β = constant is tangent to γ, a jump occurs. Although all this has been said before, it bears repeating since a similar configuration will be central in the subsequent examples.

7.3 Insects and Trees

The spruce and fir forests of eastern Canada and the northeastern United States have periodically been subject to ravages by a caterpillar called the spruce budworm. For a number of years, a given patch of forest is seen to grow with hardly any budworm in evidence. When the trees have reached a certain level of maturity there is, however, an explosive increase in the number of these insects and they begin to defoliate the trees. When a stand of mature trees have been sufficiently denuded over several consecutive years, they wither and die. The budworm population within the patch can no longer be sustained as its food supply becomes scarce. Their numbers decrease and then quite suddenly collapse to a low subsistence level. The forest canopy has now been opened up which permits new seedlings to grow. The forest renews itself and a new cycle begins, which eventually leads to a another outbreak of insects in about thirty to seventy years. Air current allow budworm moths to migrate from other forest patches, and this helps to reinstate their presence as new larvae are deposited.

Budworm growth depends on the amount of habitat available, which is measured in average branch surface area S per acre of land and on the amount of spruce and fir needles on these branches. Average budworm density B (larvae per acre of forest) is assumed to have logistic growth in which the carrying capacity is proportional to habitat size S.

There is also predation by birds and various insects and parasites, subject to satiation (it is useful, at this point, to refer to the model in Example 6.6). The maximum predation rate is b, a constant. When budworm population is small, they are not easily visable to the predators in the foliage of a growing and still healthy forest. Therefore, predation is negligible when B is small. However, as B increases the predators, which up to now has been dependent on alternate food sources, begin to concentrate their attention on the more visable prey. At the same time, overcrowding allows the parasites to be more effective.

The predation term is therefore taken to be

$$\frac{bB^2}{a^2 + B^2}$$

for some scalar a. This expression insures a negligible predation for small B, and an increasingly effective one as B increases. The predation terms is monotonely increasing in B, rising to b more and more sharply as a decreases. For an immature and growing forest, predation should be easier and so a is expected to be small; while in a fully grown and healthy stand of trees, a would of course increase because of the more difficult access to prey. In other words, a larger value of B can be sustained when S is large for the same level of predation, with the opposite being true in an immature forest with scant foliage. For this reason, a is made proportional to S: $a = k_1 S$. The equation for B is therefore

$$\dot{B} = rB\left(1 - \frac{B}{kS}\right) - \frac{bB^2}{(k_1 S)^2 + B^2}. \tag{7.17}$$

One of our goals is to show that the budworm is a resilient species in that it can survive a poor habitat. It is also unstable in the sense that its population fluctuates widely instead of remaining fairly constant.

Since the softwood evergreens of spruce and fir at important in the paper industry, the periodic outbreaks are economically disruptive. Since the 1950s, insecticides have been sprayed at a high monetary and environmental cost. As a consequence, the budworm population has been stabilized at a high endemic level where outbreak is imminent should spraying ever stop. The implications of this for forest management are significant. However, we will concern ourselves solely with modeling the budworm cycle in the absence of logging or spraying.

In the usual way, we begin by studying the equilibria of 7.17. Let us ignore the case of $B = 0$ and make a change of variables by writing B as $B = k_1 Sx$. The nontrivial equilibria of 7.17 satisfy

$$\frac{rk_1 S}{b}\left[1 - \frac{x}{(k/k_1)}\right] - \frac{x}{1 + x^2} = 0. \tag{7.18}$$

Define R as $rk_1 S/b$ and k/k_1 as the constant Q. Then 7.18 becomes

$$R\left(1 - \frac{x}{Q}\right) = g(x) \tag{7.19}$$

in which $g(x) = x/(1 + x^2)$ is a nonnegative function, which tends to zero as $x \to \infty$. It has a single maximum as well as an inflection point, as is easily verified. The left side of 7.19 is a straight line $f(x)$ of slope $-R/Q$, and the equilibria occur where this line intersects $g(x)$ as one sees in Figure 7.15.

It is fairly evident that if Q is small enough, then $f(x)$ has only one intersection with $g(x)$ for all R. However, for Q sufficiently large there can be either one or three intersections. Now R is proportional to S, the branch density of the forest, and therefore S and R are small in a newly growing patch of trees. As the forest canopy becomes more dense both S and R increase. Let us use this fact in order to follow a typical sequence of forest growth. First, however, note that 7.17 can be written as

$$a\dot{x} = x[f(x) - g(x)], \tag{7.20}$$

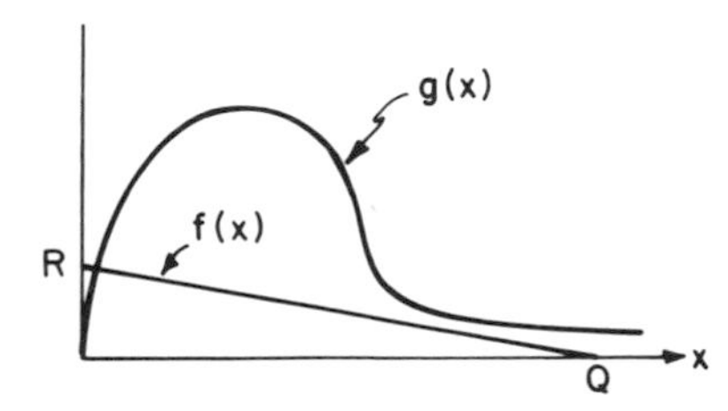

Figure 7.15 Intersection of the curve $g(x)$ in 7.19 with a straight line of slope $-R/Q$.

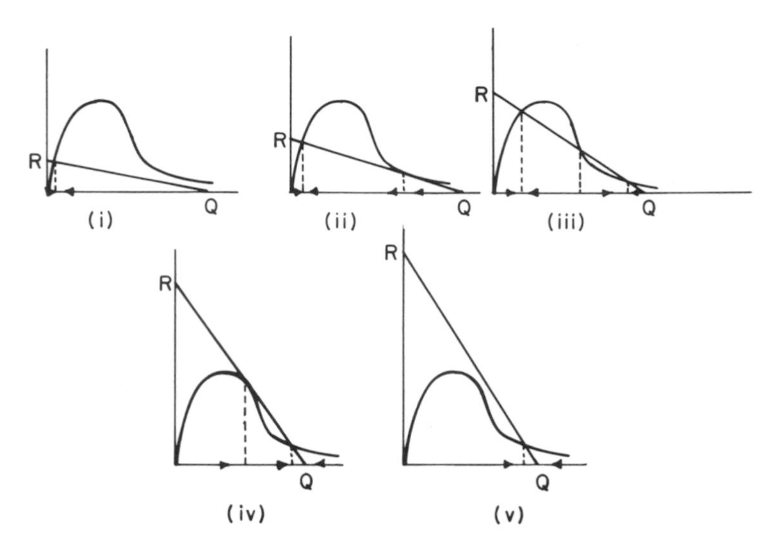

Figure 7.16 Forest growth as S and therefore R increases, with Q large. At first there is a single equilibrium (i). After a time, the line becomes tangent to the curve (ii), beyond which we find three attractors, the middle of which is repelling (iii). After another bout of tangency (iv) the equilibria coalesce into a single stable attractor (v). The stable rest point in (i) and (ii) suddenly jumps upward in (iii) and (iv) to a new equilibrium.

and so $\dot{x}$ is positive (negative) if the line is above (below) the curve $g(x)$. This reveals the stability or instability of the equilibria, as shown by the arrows in the figure. Also, note that since x increases with S so does B.

By following Figure 7.16 backward, it is possible to envisage the decline of the forest as R decreases. We see that a stable equilibrium in (v) jumps down to a lower level in (i) after tangency occurs in (ii). Thus, the budworm essentially resides at essentially one or two extreme states, subsistence and outbreak with rapid jumps between. It is therefore bistable (Figure 7.17).

Figure 7.17 makes us surmise that a cusp catastrophe looms in the background. Indeed, 7.18 and 7.19 define the zeros of U where U is a smooth potential function that depends on two parameters R and Q. This is seen more clearly by multiplying 7.18 by $1 + x^2$ to get

$$-x^3 + Qx^2 - \left(1 + \frac{Q}{R}\right)x + Q = 0. \tag{7.21}$$

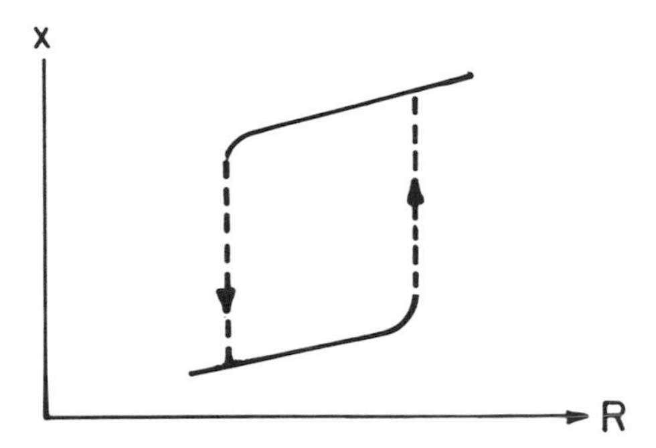

Figure 7.17 Budworm cycle as forest density grows and then declines.

The potential function is then

$$U(x, R, Q) = -\frac{x^4}{4} + \frac{Qx^3}{3} - \frac{1}{2}\left(1 + \frac{Q}{R}\right)x^2 + Qx. \quad (7.22)$$

Theorem 7.1 leads us to expect that U can be smoothly deformed into the standard cusp form. This can be shown directly by letting $x = y + Q/3$. Substituting into 7.21 gives a new equation in y in which the quadratic term has been eliminated:

$$-y^3 + \alpha y + \beta = 0, \quad (7.23)$$

where the parameters α and β depend on R and Q. Thus, the cusp model is obtained by a smooth change of variables. The jumps in x occur where the line $f(x)$ is tangent to $g(x)$. This means that the catastrophe set is defined by the following two equations:

$$R\left(1 - \frac{x}{Q}\right) = \frac{x}{1 + x^2}$$

and (7.24)

$$-\frac{R}{Q} = \frac{1 - x^2}{(1 + x^2)^2}.$$

A bit of algebra is needed to show that R and Q lie on a roughly crescent shaped curve in the positive quadrant of the R, Q plane. The map $x = y + Q/3$ transforms this curve into the standard cusp. The

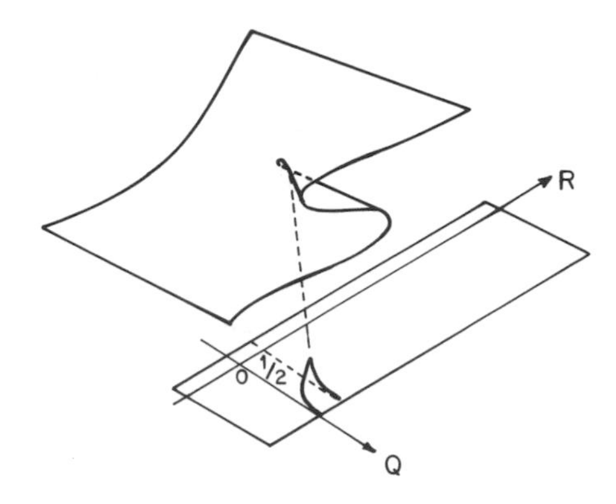

Figure 7.18 Cusp catastrophe model of budworm growth. The upper sheet represents an outbreak level while the lower is a subsistence level. Note the similarity to Figure 7.4.

surface defined implicitly by 7.19 is displayed in Figure 7.18 and is a distorted version of the standard model in Figure 7.4.

At this juncture, it is necessary to introduce certain feedback links in order to more fully describe a budworm cycle. Let E denote the percentage of foliage on the trees, with E close to unity indicating a healthy forest. Once the trees are attacked by the budworm, E begins to decline toward zero. Tree growth is stunted if the needles are gone and eventually branches and trees die with continued defoliation. We assume logistic growth for the foliage minus the effects of budworm stress, which we take to be proportional to average budworm density per average branch size, namely B/S. In order to insure that B/S tends to zero as B decreases (to simulate diminishing stress), we multiply B/S by E^2. The equation for E can then be written as

$$\dot{E} = r_E E(1 - E) - \frac{pBE^2}{S} \tag{7.25}$$

in which p is some constant of proportionality.

Once defoliation has begun in a mature forest, the average branch size is still large for a period of time. But with continued stress, the branches and trees ultimately are stunted and die. Therefore, the carrying capacity kS of the budworm larvae is eventually curtailed as E decreases. The equation for B should be modified to account for this decline. We do this by changing the carrying capacity of kSE and redefining Q to be kE/k_1. Finally, allow S itself to grow logistically with a maximum capacity proportional to E:

$$\dot{S} = r_S S\left(1 - \frac{S}{k_S E}\right). \tag{7.26}$$

Pick a value of B at either a high or low level. With B now fixed, we allow S and E to reach equilibrium with each other. Setting the righthand sides of 7.25 and 7.26 to zero, we see that

$$S = k_S E \tag{7.27}$$

and

$$S = \frac{pEB}{r_E(1 - E)}. \tag{7.28}$$

Isocline analysis reveals that the origin is unstable, but the nontrivial rest point is a.s. (Figure 7.19). As B increases, this equilibrium tends to zero. Let us follow a typical scenario. Initially in a growing and healthy forest, B is small. E and S slowly attain an equilibrium in which E is close to unity and S is also large. The movement is slow since it takes place over a number of years. As this happens B eventually jumps rapidly (within the time scale being considered here) to a larger value in response to the

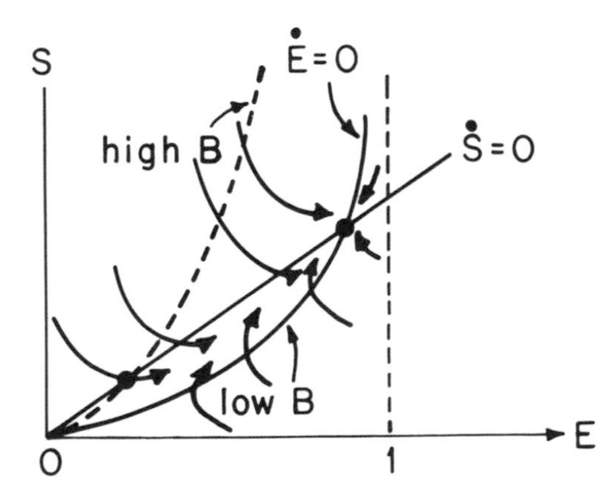

Figure 7.19 Isocline analysis showing the relation between S and E in Equation 7.25 and 7.26 for high and low B values.

changing values in forest density. As a consequence, both E and S decline, slowly, to a much lower stable equilibrium. Since R and Q are proportional to S and Q, respectively, it follows that these two parameters also decline and so B is forced to jump to a lower value at some point. S and E and therefore R and Q now renew their move to a higher equilibrium and the process repeats itself. This is the budworm cycle.

The differential equations for S and E are, in effect, the feedback equations for R and Q, linking these parameters to B.

A small technical point needs to be made at this point which is that the isoclines in Figure 7.19 have a unique nontrivial intersection for all B provided the slope of the $\dot{E} = 0$ isocline is less than that of the line $\dot{S} = 0$ at $E = 0$. But this is easily achieved if p is small enough. Finally, since the catastrophic jump in B depends on the fact that $f(x)$ can have three intersections with $g(x)$, it is reasonable to ask whether Q ever gets large enough to permit this to happen. Field data indicates that indeed this is the case and so the model does generally conform to what is actually observed.

7.4 The Earth's Magnet

This section assumes an elementary knowledge of electric and magnetic fields as taught in an introductory physics course. Our goal is to develop a plausible model for the magnetic field reversals of the earth. Such pole reversals have taken place in an irregular manner an average of every 340,000 years during the last 40 million years. Between reversals the field fluctuates slightly but in an apparently random way. The reversal time is relatively fast and requires only a few thousand years.

Consider a conducting disc of radius R which rotates at an angular speed $\omega = \dot{\theta}$ about a vertical axis. A conducting coil is connected to the axle and to the disc periphery as shown in Figure 7.20. The whole apparatus is known as a dynamo.

The moment of inertia of the disc is C. We start the disc at some speed ω_0 and the coil has an initial current I_0. A constant torque G is applied to the axle to turn the disc. This represents the constant rotation of the earth. We imagine that the earth's molten core sets up a magnetic field as it moves, which corresponds to a magnetic field induced by a current I in the coil. The analogy between the earth's core and a dynamo

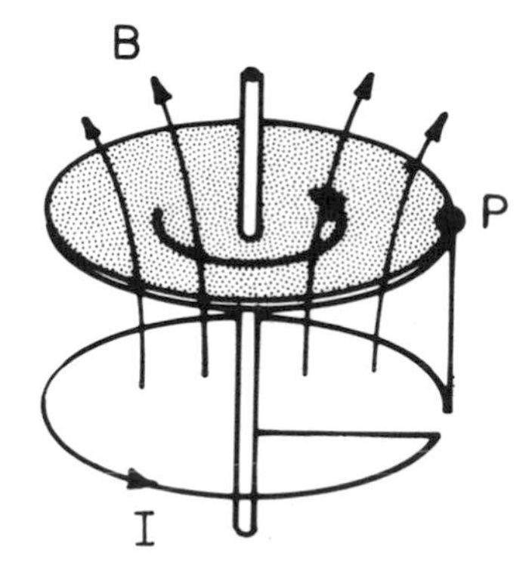

Figure 7.20 Schematic of a dynamo.

is admittably crude and is meant only to be suggestive of what is going on.

If L is the self-inductance and R the resistance in the R, L circuit then the voltage V is, as is well known,

$$V = RI + L\dot{I}.$$

Current in the coil sets up a magnetic field B which cuts the disc in a perpendicular manner. As the disc moves, the field induces a current to flow radially outward from the axle to the periphery (from 0 to P in the figure). As the disc rotates through an angle θ, the magnetic flux along the radial direction is

$$\phi = \theta \int_0^T Br\,dr = b\theta,$$

where b is the value of the integral. The potential difference between 0 and P is therefore $V = b\dot{\theta} = b\omega$. It follows that

$$L\dot{I} + RI = b\dot{\omega}. \tag{7.29}$$

The rotational torque of the disc is $C\omega$, which equals the constant applied torque G minus viscous damping, which is proportional to ω minus the opposing magnetic torque. The last term is derived as follows. The field B, normal to the disc, acts on a radial element of length Δr with a force given by $I\Delta r \times B = IB\Delta r$ in a direction opposite that of rotation. The radial element is a distance r from the axle and so the total magnetic torque is

$$I \int_0^R Br\,dr = Ib.$$

Thus,

$$C\dot{\omega} = G - k\omega - Ib, \tag{7.30}$$

for some positive constant k. We know that b varies with I. Therefore, b also changes with I, and we can write b as $b = MI$ in which the proportionality constant M is mutual inductance. It follows that 7.29 and 7.30 may be expressed as

$$\begin{aligned} L\dot{I} &= \omega MI - RI \\ C\dot{\omega} &= G - I^2M - k\omega. \end{aligned} \tag{7.31}$$

Let us now add a small magnetic field normal to the disc to represent the influence of all external sources. This *ad hoc* assumption is removed in the next chapter where an alternate formulation of the model is established that gives results qualitatively similar to the present one.

The additional field gives rise to a current I_1 and therefore 7.31 is modified to give

$$\begin{aligned} L\dot{I} &= M\omega(I + I_1) - RI \\ C\dot{\omega} &= G - M(I + I_1)I - k\omega. \end{aligned} \tag{7.32}$$

It is convenient to rewrite these equations by letting

$$I_1 = \sqrt{\frac{G}{M}}\,\eta$$

$$I = \sqrt{\frac{G}{M}}\,x_1$$

$$\omega = \sqrt{\frac{GL}{CM}}\,x_2$$

$$t = \sqrt{\frac{CL}{GM}}\,\tau$$

$$\mu = R\sqrt{\frac{C}{GLM}}$$

$$\nu = k\sqrt{\frac{L}{GCM}}\,.$$

Then, with time now parametrized by τ, 7.32 becomes

$$\begin{aligned}\dot{x}_1 &= (x_1 + \eta)x_2 - \mu x_1 \\ \dot{x}_2 &= 1 - x_1(x_1 + \eta) - \nu x_2.\end{aligned} \tag{7.33}$$

The equilibria are found by multiplying the equation $\dot{x}_1 = 0$ by ν and then substituting into the relation $\dot{x}_2 = 0$. This gives a cubic equation in x_1:

$$-x_1^3 + (1 - \eta^2 - \mu\nu)x_1 - 2\eta x_1^2 + \eta = 0. \tag{7.34}$$

By a smooth change of variables $x = y - 2\eta/3$, the quadratic term is eliminated and we obtain

$$-y^3 + \alpha y + \beta = 0. \tag{7.35}$$

This cubic equation is identical to relation 7.6 that defines the standard cusp catastrophe surface $\mathscr{M}$. The parameters α and β are dependent on η and $\mu\nu$. Thus, x_1 is a root of $U' = 0$ where U is a smooth potential function in the parameters η and $\mu\nu$. The surface defined by 7.34 is smoothly deformable to $\mathscr{M}$. Since I, and therefore B, are proportional to x_1, it follows that the magnetic field attains one of two bistable equilibrium states on a cusp catastrophe surface. To see more clearly what is going on let $x = x_1$ and rewrite 7.34 as

$$f(x) = g(x)$$

in which $f(x)$ is the straight line $x(1 - \mu\nu) + \eta$ and $g(x)$ is the cubic curve $x(x + \eta)^2$. The intersection of f and g is shown in Figure 7.21 for $\mu\nu < 1$. When $\mu\nu$ is greater than one, the slope of the line is negative, and there is but a single real root. This is apparent from the figure. For three real roots to exist, we therefore require that $\mu\nu$ be less than unity.

We know from the discussion in Section 7.1 and from the comments at the end of Section 7.2 that the cubic $-y^3 + \alpha y + \beta = 0$ has three real roots if β, and therefore η, is small enough. In fact, $\beta = \eta(1 + 2\mu\nu)/3 + 2\eta^3/27$. This can also be seen directly (Exercise 7.5.5).

The external field has an intensity measured by η. Suppose η fluctuates at random about a mean of zero with small amplitudes (that is, with $|\eta|$ small), and let us follow what happens in Figure 7.22 as it crosses from positive to negative values.

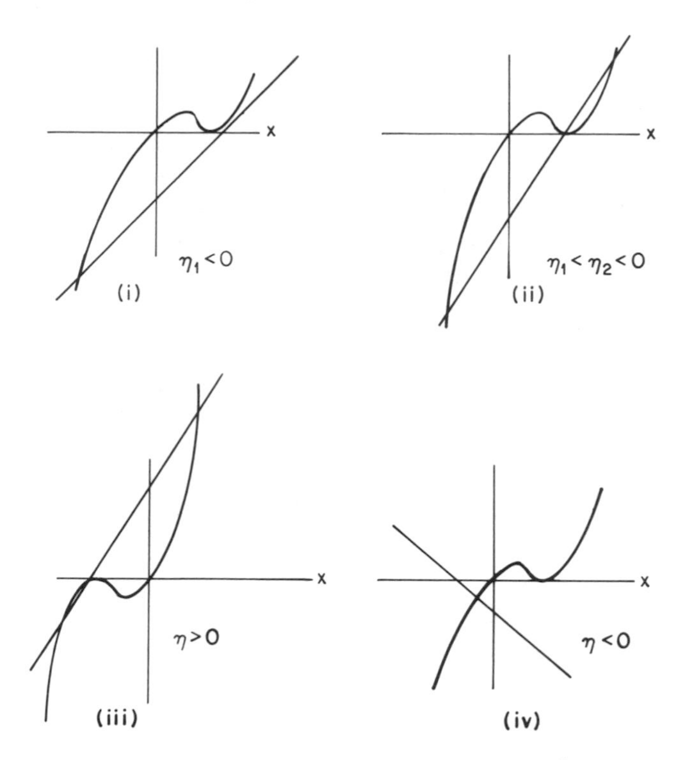

Figure 7.21 Intersection of the line $x(1-\mu\nu)+\eta$ and the curve $x(x+\eta)^2$ for different η. This is shown for $\mu\nu<1$ in (i) to (iii), and one sees that there are three intersections for $|\eta|$ small enough. In (iv) only one intersection is possible. This corresponds to $\mu\nu>1$.

The magnetic field B is proportional to x_1, as we have seen. As η varies, B is assumed to reach its equilibrium immediately. This is represented by the intersection of the line and the curve in the figure above. We see that it changes polarity by jumping to a new equilibrium as η passes a certain threshold. With η moving in reverse, another jump occurs in the opposite direction but along a hysteresis path (Figure 7.23). We have assumed $\mu\nu<1$. Otherwise, B varies with η but does not change polarity. The combined effect is exhibited as a cusp surface in Figure 7.24.

The model therefore displays the possibilities of field reversals that occur erratically depending on the vagaries of small external magnetic disturbances, which can suddenly flip B to a new polarity. Between flips

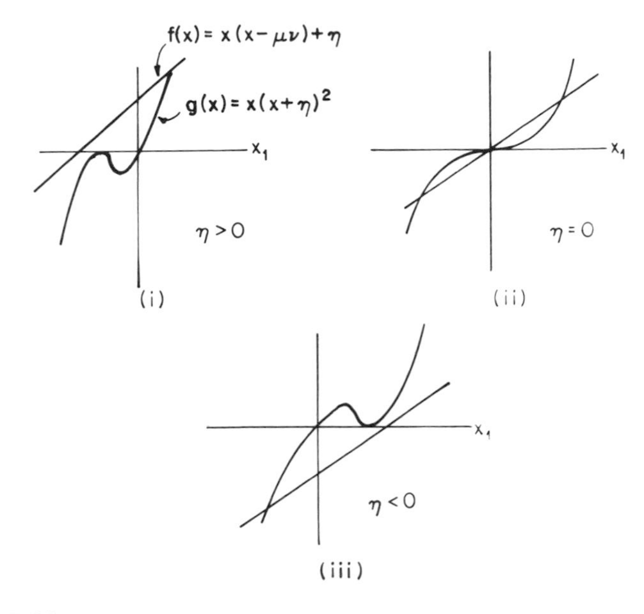

Figure 7.22 A typical scenario of how the current I and therefore the magnetic field B (as represented by the variable x_1) vary with η, as η decreases from positive to negative values. In (i) there is a single positive equilibrium at which the field has one of its possible polarities. As η decreases, the line reaches tangency with the curve giving birth to new equilibria by bifurcation (ii). β has decreased slightly but retains its polarity. Eventually, the line is again tangent with the curve and the equilibria coalesce. Beyond this, there is again a single rest point (iii) and B has jumped to an opposite polarity.

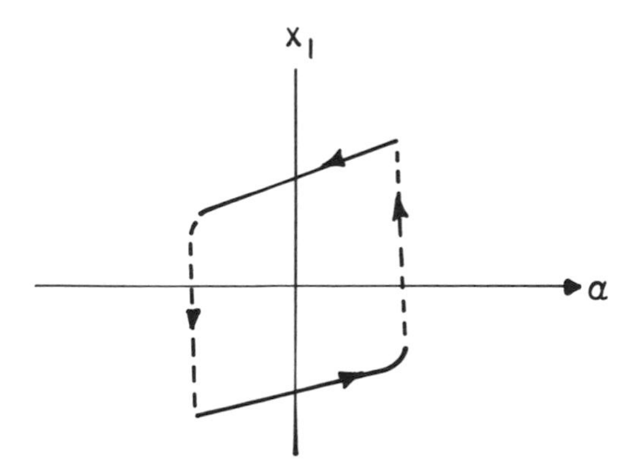

Figure 7.23 The change in magnetic field, represented by x_1, as a hysteresis curve.

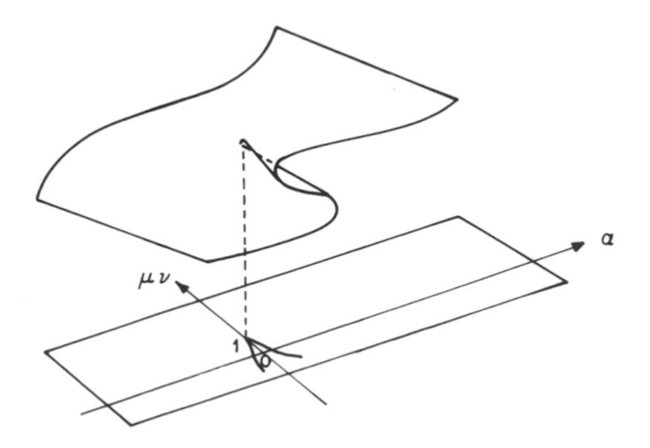

Figure 7.24 Cusp catastrophe model of the earth's magnetic field. The upper and lower sheets represent different polarities. Note the similarity to Figure 7.4.

it lies on one sheet or the other of a cusp surface where it shifts about a bit as η varies at random.

7.5 Exercises

7.5.1 We recall from Section 5.4 that traffic density ρ has the property that, whenever traffic is slow on the highway ahead, the fast moving cars eventually catch up with the slow. Mathematically, this situation permits the fast traffic to pass the slow thereby creating a wave-like density that is multivalued (Figure 5.9). This occurs in a region in which characteristics intersect. Physically, of course, a multivalued density is not possible, and so when fast traffic catches up with slow, a shock wave is created.

There is an idea here that is suggestive of a cusp catastrophe. Suppose that for each x and t, the density ρ minimizes some unknown potential function U. Then one would expect a cusp catastrophe to emerge in terms of the parameters x and t. The pleat in the catastrophe surface would correspond to the (hypothetical) wave-like density whenever characteristics intersect. Draw a plausible representation of this phenomenon allowing for an initial density that increases with x (slow traffic ahead).

7.5.2 Discuss the similarities and differences between the model proposed in Exercise 6.5.12 and the Zeeman model of Section 7.2.

7.5.3 Show that Equation 7.13 and 7.15 can be written in the form

$$\dot{x} = \nu\left[\beta - \left(\frac{x^3}{3} - \alpha x\right)\right]$$

$$\dot{\beta} = x - \bar{x},$$

for ν a large positive constant, simply by changing the time scale t to some suitable τ.

7.5.4 Reconsider the fishery model of Section 6.5 in which

$$N = f(N) - \nu EN \tag{7.36}$$

and $f(N)$ has the form shown in Figure 4.2. The equilibria of 7.36 occur at the intersection of $f(N)$ with the straight line $g(N) = \nu EN$. To makes matters more interesting, modify the harvesting term to $\nu EN - \mu$ for μ positive. μ represents an externally imposed ban on fishing: when N is small enough harvesting becomes zero. Therefore, μ is a surrogate for environmental legislation. It follows that, depending on the values of μ and E, there are either one or three equilibria as we see from the figure below.

Show that one obtains a catastrophe model in which overharvesting leads to a sudden collapse of the fish population to near

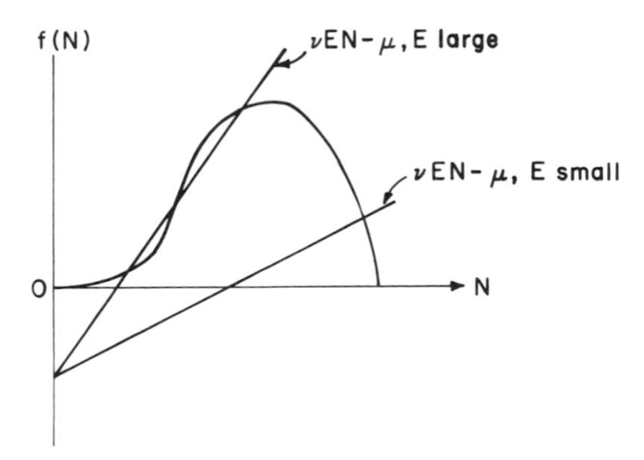

Figure 7.25 A fish harvesting model. The curve represents autocatalytic growth for $N < \hat{N}$, and the line is defined by $\nu EN - \mu$.

zero levels. Introduce a feedback linkage in terms of a differential equation for one of the parameters μ or E so that a cycle ensues for the fish stocks (refer to the discussion in Example 6.7). What happens as μ tends to zero or if μ is very large?

7.5.5 Rewrite the cubic relation 7.34 as

$$-x^3 + (1 - \mu\nu)x = 2\eta x^2 + \eta^2 x - \eta$$

in which $x = x_1$ and $\mu\nu < 1$. Graph the two sides of this equation and establish that there are three real roots only if $|\eta|$ is small enough.

7.5.6 In Exercise 6.5.8, we modeled algae "blooms". Such blooms represent the explosive growth of algae that occur from time to time, usually in the summer months, in coastal waters and lakes. The model related to algae population to that of tiny crustaceans that graze on the algae and to the quantity of nutrients that the algae intake.

An alternate model follows the reasoning used in studying the spruce budworm outbreak in Section 7.3 where there is also a sudden and explosive increase in population. Find a differential equation for algae density (average number per unit volume of water) in a bay having ocean access. Assume logistic growth with a carrying capacity that depends on the parameter σ/Q. σ is the average input rate of nutrients into the bay (amount per volume of water), and Q is the average rate of outflow of bay water into the ocean (volume of water per unit time). An increase in σ is an inducement to growth, whereas an increase in Q restricts the habitat since both nutrient and algae are swept out into the ocean in each ebb tide, never to return. Ocean water that returns in a flood tide is assumed to be free of nutrients and algae. In effect, a high value of σ/Q indicates a favorable nutrient concentration and low tidal dispersion.

Predation (namely, grazing by copepods and other organisms) is inefficient at low algae levels but becomes very effective at high levels. Indeed, a high algae density encourages its own self-destruction by becoming toxic to itself. Per-capita growth is cyclic to account for the annual change in sunlight and temperature, each of which stimulates algae growth.

Develop a catastrophe model and argue why blooms occur in the summer months if σ/Q is large enough. Such bursts of growth are undesirable because they absorb oxygen from the water rendering it unfit for swimming and unhealthy to fish. Moreover, the algae release substances toxic to shellfish, and when the algae die in large numbers, they leave a green slime along the beaches.

The model assumes spatial homogeneity. You should refer to Sections 5.2 and 5.3 for related models in which spatial factors are explicitly incorporated.

Chapter **Eight**

Chaos

8.1 Not All Attractors Are Limit Cycles or Equilibria

Our analysis of the long term behavior of orbits has been restricted to limit sets that are either equilibria or closed orbits. In planar systems, this is essentially all that can occur by virtue of Theorem 6.1 (Poincaré–Bendixson). For gradient systems, moreover, only equilibria are possible. In general, however, other kinds of limiting behavior may appear, some of it quite complicated, for dynamical systems in $\mathbf{R}^k$, $k \geq 1$. In this chapter, we want to introduce some of the simpler examples of this type to illustrate what can happen. An indispensible part of this study is the numerical computation of orbits. Such computations reveal something of what may be expected in the long run, as time increases, especially since more precise mathematical results are often lacking.

Let us consider the vector system

$$\dot{\mathbf{x}} = \mathbf{f}(\mathbf{x}) \tag{8.1}$$

with $\mathbf{x}$ in some open set U of $\mathbf{R}^k$. The positive limiting set γ^+ of an orbit γ of 8.1 was defined in Section 6.1. We repeat it here for convenience.

Let $\mathbf{x}(t)$ be a solution of 8.1 in U. Then γ^+ is the set of points $\mathbf{p}$ for which

$$\lim_{n \to \infty} \|\mathbf{x}(t_n) - \mathbf{p}\| = 0$$

as $t_n \to \infty$ through some sequence of values. The two examples that come immediately to mind are the a.s. rest points and limit cycles, both of which have been studied extensively in this book. Now we wish to enlarge on this concept by defining a set Λ in U to be an *attractor* if there is some set Ω_0 containing Λ such that the positive limiting set of any orbit that begins in Ω_0 is contained in Λ. Thus, the origin is a point attractor in $\mathbf{R}^2$ for the damped harmonic oscillator, while the unique limit cycle of the Van der Pol oscillator is an attractor in the set consisting of the plane minus the origin. In general, however, Λ is not a simple set and the motion of an orbit near it may appear to be chaotic. The largest possible Ω_0 is called the *basin of attraction* of Λ.

Let us restrict our dynamical systems to $\mathbf{R}^3$. Consider a bounded set B having a smooth and closed boundary S. If all orbits that cross S do so in an inward direction, then B is called a *trapping region*. In effect, B is invariant in the sense that an orbit that begins in B remains there for all future time.

Suppose, further, that the divergence

$$\sum_{i=1}^{3} \frac{\partial f_i}{\partial x_i}$$

of the vector field $\mathbf{f}$ in 8.1 is strictly negative. Under these conditions, we wish to show that a trapping region B contains an attractor Λ having zero volume. Thus, Λ is a "thin" set in $\mathbf{R}^3$. However, its appearance is not obvious, and it certainly need not be a single point or a closed cycle, as the examples will show.

To establish the assertion made above consider all orbits that begin in B at some fixed time t. They wander about in B and at some later time $t + s$ are found in some new set $\Phi_s(B)$ that is the image of B under the transformation 8.1. $\Phi_s(B)$ is found by integrating 8.1 forward from all possible initial values in B from time t to $t + s$. This results in a uniquely defined vector

$$\Phi_s(\mathbf{x}(t)) = \mathbf{x}(t + s) = \mathbf{x}(t) + \int_t^{t+s} \mathbf{f}(\mathbf{x}(\tau))\, d\tau. \tag{8.2}$$

The integral is the vector with components

$$\int_t^{t+s} f_i(\mathbf{x}(\tau))\, d\tau, \qquad \text{for } 1 \le i \le 3.$$

Let $V(t)$ be the volume of B at time t and $V(t+s)$ the volume of the image set $\Phi_s(B)$. Denote by J_s the determinant of the Jacobian matrix of the transformation $\Phi_s(B)$. The Jacobian matrix, evaluated at $\mathbf{x}(t)$, has components

$$\frac{\partial x_j(t+s)}{\partial x_i(t)}, \qquad \text{for } 1 \le i,\ j \le 3.$$

Thus,

$$J_s = \text{Det}\left(\frac{\partial \Phi_s(\mathbf{x}(t))}{\partial \mathbf{x}(t)}\right).$$

Applying the first-order Taylor approximation theorem to each scalar component of the integral in 8.2 gives

$$\Phi_s(\mathbf{x}(t)) = \mathbf{x}(t) + s\mathbf{f}(x(t)) + o(s).$$

Therefore,

$$\frac{\partial \Phi_s(\mathbf{x}(t))}{\partial \mathbf{x}(t)} = I + sA + o(s)$$

in which I is the identity matrix and A is the Jacobian matrix of $\mathbf{f}$ evaluated at $\mathbf{x}(t)$. A has components

$$a_{ij} = \frac{\partial f_i(\mathbf{x}(t))}{\partial x_j}, \qquad 1 \le i,\ j \le 3.$$

Thus,

$$J_s = \text{Det}\begin{pmatrix} 1 + sa_{11} & sa_{12} & sa_{13} \\ sa_{21} & 1 + sa_{22} & sa_{23} \\ sa_{31} & sa_{32} & 1 + sa_{33} \end{pmatrix} + o(s)$$

$$= 1 + s(a_{11} + a_{22} + a_{33}) + o(s)$$

or

$$J_s = 1 + s\,\text{Trace}\,A + o(s). \tag{8.3}$$

By the change of variables formula in calculus,

$$V(t+s) = \iiint_{\Phi_s(B)} dx_1\, dx_2\, dx_3 = \iiint_B J_s\, dx_1\, dx_2\, dx_3.$$

After differentiating under the integral sign we obtain from 8.3 that

$$\frac{dV(t)}{dt} = \lim_{s\to 0} \frac{V(t+s)-V(s)}{s} = \iiint_B \text{Trace}\, A\, dx_1\, dx_2\, dx_3$$

$$= \iiint_B \text{Div}\,\mathbf{f}\, dx_1\, dx_2\, dx_3.$$

Assume now that Div $\mathbf{f}$ is some negative constant. It follows that $\dot{V} = (\text{Div}\,\mathbf{f})V$. Integrating, we obtain

$$V(t) = \text{const} \cdot e^{(\text{Div}\,\mathbf{f})t},$$

which tends to zero as t increases.

Since volumes shrink to zero, it is fairly apparent that the orbits eventually reside in the set Λ defined by

$$\Lambda = \bigcap_{s\geq 0} \Phi_s(B). \tag{8.4}$$

Therefore, Λ is an attractor and B lies in the basin of attraction. There is something reminiscent here of the Liapunov Theorem for a.s. point attractors. In fact, the volume V is a sort of Liapunov function since $\dot{V}$ is negative.

It is known, incidently, that volume is preserved in conservative dynamical systems. This is most clearly seen in the case of a one dimensional conservative equation of the form $\ddot{p} + f(p) = 0$. Written as a first-order system $\dot{x}_1 = x_2$ and $\dot{x}_2 = -f(x_1)$, it is clear that the vector field so defined has zero divergence. Therefore, $V(t)$ is constant and volume is preserved, which is a statement of Liouville's Theorem in mechanics.

Example 8.1 (Competition) In Chapter Four, we studied the dynamics of competition between two species. The resulting quadratic model in $\mathbf{R}^2$

offered few surprises. However, in $\mathbf{R}^3$ a different picture emerges. Let $N_i(t)$ be the populations of three competitors C_i, $i = 1, 2, 3$. To simplify matters, we make the convenient assumptions that each competitor has the same growth rate and carrying capacity (each of which is taken to be unity). Moreover, the competitive advantage C_2 over C_1 is the same as that of C_3 over C_2 and C_1 over C_3. Similarly, the advantage of C_3 over C_1 is identical to that of C_1 over C_2 and C_2 over C_3. This leads to the equations

$$\begin{aligned} \dot{N}_1 &= N_1(1 - N_1 - \alpha N_2 - \beta N_3) \\ \dot{N}_2 &= N_2(1 - \beta N_1 - N_2 - \alpha N_3) \\ \dot{N}_3 &= N_3(1 - \alpha N_1 - \beta N_2 - N_3) \end{aligned} \tag{8.5}$$

in which α and β are positive constants.

Other than the equilibria in which one or more competitors are zero, there is a single nontrivial equilibrium given by

$$\overline{N} = \frac{1}{1 + \alpha + \beta} \begin{pmatrix} 1 \\ 1 \\ 1 \end{pmatrix}. \tag{8.6}$$

The Jacobian matrix of the linearized system corresponding to 8.5 is easily computed to be

$$A = \begin{pmatrix} 1 - 2\overline{N}_1 - \alpha\overline{N}_2 - \beta\overline{N}_3 & -\alpha\overline{N}_1 & -\beta\overline{N}_1 \\ -\beta\overline{N}_2 & 1 - 2\overline{N}_2 - \alpha\overline{N}_3 - \beta\overline{N}_1 & -\alpha\overline{N}_2 \\ -\alpha\overline{N}_3 & -\beta\overline{N}_3 & 1 - 2\overline{N}_3 - \alpha\overline{N}_1 - \beta\overline{N}_2 \end{pmatrix}, \tag{8.7}$$

and so at the equilibrium 8.6, we obtain

$$A = -\frac{1}{1 + \alpha + \beta} \begin{pmatrix} 1 & \alpha & \beta \\ \beta & 1 & \alpha \\ \alpha & \beta & 1 \end{pmatrix}. \tag{8.8}$$

It follows that $\overline{N}$ is a.s. if the eigenvalues of

$$M = \begin{pmatrix} 1 & \alpha & \beta \\ \beta & 1 & \alpha \\ \alpha & \beta & 1 \end{pmatrix} \tag{8.9}$$

have positive real parts. The characteristic equation of M is

$$\mathrm{Det}(M - \lambda I) = (1 - \lambda)^3 + \alpha^3 + \beta^3 - 3\alpha\beta(1 - \lambda).$$

Let $u = \lambda - 1$. Then we need to find the roots of the cubic equation

$$P(u) = u^3 - 3\alpha\beta u - (\alpha^3 + \beta^3). \tag{8.10}$$

One root is $u_1 = \alpha + \beta$, as a quick check shows. Hence, 8.10 becomes

$$P(u) = (u - (\alpha + \beta))(u^2 + au + b).$$

Expanding this expression and comparing terms with 8.10 shows that $a = \alpha + \beta$ and $b = (\alpha + \beta)^2 - 3\alpha\beta$. Therefore, the problem reduces to finding the roots of the quadratic equation

$$u^2 + (\alpha + \beta)u + (\alpha + \beta)^2 - 3\alpha\beta = 0.$$

These are easily obtained as

$$u_{2,3} = -\frac{\alpha + \beta}{2} \pm \frac{i\sqrt{3}(\alpha - \beta)}{2},$$

and so the three eigenvalues of M are

$$\begin{aligned} \lambda_1 &= 1 + \alpha + \beta \\ \lambda_{2,3} &= 1 - \frac{\alpha + \beta}{2} \pm \frac{i\sqrt{3}(\alpha - \beta)}{2}. \end{aligned} \tag{8.11}$$

Since α and β are positive, the eigenvalues have positive real parts if $\alpha + \beta < 2$. When $\alpha + \beta > 2$, there is instability.

Observe that if two of the species, say N_2 and N_3, are zero then the matrix A becomes, at the rest point $\begin{pmatrix} 1 \\ 0 \\ 0 \end{pmatrix}$,

$$\begin{pmatrix} -1 & -\alpha & -\beta \\ 0 & 1 - \beta & 0 \\ 0 & 0 & 1 - \alpha \end{pmatrix}.$$

Its eigenvalues satisfy the cubic

$$(1 + \lambda)[1 - (\beta + \lambda)][1 - (\alpha + \lambda)] = 0$$

and are therefore -1, $1 - \alpha$, and $1 - \beta$. It follows that $\begin{pmatrix} 1 \\ 0 \\ 0 \end{pmatrix}$ is unstable when either $\alpha < 1$ or $\beta < 1$ or both. A similar argument establishes that the other single species equilibria $\begin{pmatrix} 0 \\ 1 \\ 0 \end{pmatrix}$ and $\begin{pmatrix} 0 \\ 0 \\ 1 \end{pmatrix}$ are also unstable. From now on, we restrict ourselves to the unstable three-species equilibrium in which $\alpha + \beta > 2$ and $\alpha < 1$.

Denote the sum $\Sigma_{i=1}^{3} N_i$ by S. From 8.5 one then obtains the equation

$$\dot{S} = S(1 - S) + \gamma(N_1 N_2 + N_2 N_3 + N_3 N_1), \tag{8.12}$$

where $\gamma = \alpha + \beta - 2$. Also, if $P = N_1 N_2 N_3$, then

$$\frac{d \ln P}{dt} = -\gamma + (3 + \gamma)(1 - S). \tag{8.13}$$

At this juncture, further progress requires that 8.12 be integrated numerically. What is found is that the orbits asymptotically approach the corners of the triangular region defined by $\Sigma_{i=1}^{3} N_i = 1$ and $N_i > 0$ in a cyclic fashion. Moreover, the orbits appear to linger more and more about the vertices of the triangle as they get closer, with increasingly rapid movements from corner to corner (Exercise 8.4.3). It is therefore approximately true that $N_1 N_2 \sim 0$, $N_2 N_3 \sim 0$, and $N_3 N_1 \sim 0$ most of the time. The Equations 8.12 and 8.13 are consequently approximated roughly by

$$\dot{S} = S(1 - S)$$

$$\frac{d \ln P}{dt} = -\gamma.$$

It follows that as t increases, the sum $S(t)$ goes to one, while the product $P(t)$ moves to a constant. Thus, all orbits approach the plane $\Sigma_{i=1}^{3} N_i = 1$ in the positive orthant. However, the corner points of this triangular region are unstable equilibria, and so even though these points belong to the attracting set Λ, the set Λ itself cannot be a closed orbit. Moreover, all other orbits loop about with longer and longer periods as they get closer to Λ. We conclude that Λ is a non-periodic attractor. This represents a situation in which C_1 is dominant for awhile (with C_2

and C_3 nearly extinct) and then, quite suddenly, C_2 is dominant, then C_3, and so on.

It is worth noting that the vector field **f** in 8.5 satisfies the divergence condition $\text{Div}\,\mathbf{f} < 0$ when t is large enough (Exercise 8.4.1). However, since we already know that the orbits ultimately reside in a plane, the fact that the attractor has zero volume in $\mathbf{R}^3$ is immediate in this case.

8.2 Strange Attractors

In Section 7.4, a catastrophe model was devised to suggest a mechanism for the reversal of the earth's magnetic field. Let us now approach this problem in a different way by working with a modified dynamo configuration as shown in Figure 8.1. We retain the notation and terminology used in Section 7.4 and suggest that you review the derivation given there before proceeding.

What one now has are two rotating discs and coils connected so that the current in one coil feeds the magnetic field of the other. Each circuit has the same self-inductance L and resistance R, and the discs are driven by the same constant torque G.

The currents in the two circuits are I_1 and I_2, and the voltage generated in each depends now on the magnetic field set up by the current in the other. Let the angular velocities of the discs be ω_1 and ω_2.

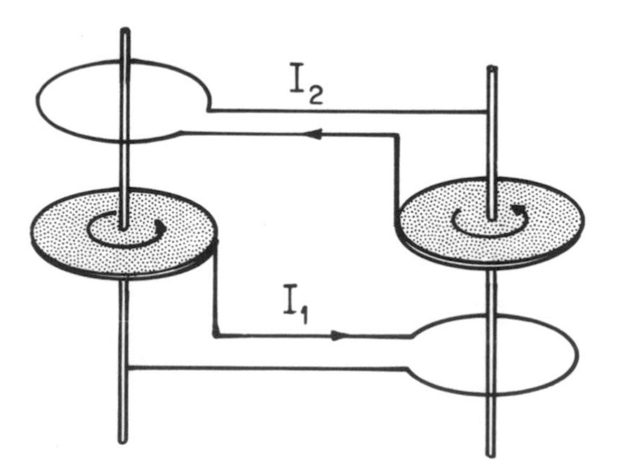

Figure 8.1 A modified dynamo configuration.

If M is the mutual inductance between circuits, then, analogously to the way Equations 7.31 were derived, one obtains

$$\begin{aligned} RI_1 + L\dot{I}_1 &= MI_2\omega_1 \\ RI_2 + L\dot{I}_2 &= MI_1\omega_2. \end{aligned} \tag{8.14}$$

The rotational torque in each disc is $I\dot{\omega}$ and the opposing torque is, by an argument similar to that used earlier, MI_1I_2. This time, however, viscous damping is ignored. Thus,

$$\begin{aligned} C\dot{\omega}_1 &= G - MI_1I_2 \\ C\dot{\omega}_2 &= G - MI_1I_2. \end{aligned} \tag{8.15}$$

From 8.15, it is seen that $\omega_1 - \omega_2$ is a constant, which we write as

$$\omega_1 - \omega_2 = \sqrt{\frac{GL}{CM}}\,\gamma \tag{8.16}$$

for some γ. Let $I_i = \sqrt{G/M}\,x_i$, $\omega_i = \sqrt{GL/CM}\,y_i$, and $t = \sqrt{CL/GM}\,\tau$. Then 8.14 becomes

$$\begin{aligned} \dot{x}_1 &= y_1x_2 - \mu x_1 \\ \dot{x}_2 &= y_2x_1 - \mu x_2 \\ \dot{y}_1 &= \dot{y}_2 = 1 - x_1x_2, \end{aligned} \tag{8.17}$$

where $\mu = (R/L)\sqrt{LC/MG}$. Note that because of 8.16 and the relation

$$\omega_1 - \omega_2 = \sqrt{\frac{GL}{CM}}\,(y_1 - y_2),$$

we obtain

$$\gamma = y_1 - y_2.$$

Let $y = y_1$. Then 8.17 can be rewritten as

$$\begin{aligned} \dot{x}_1 &= yx_2 - \mu x_1 \\ \dot{x}_2 &= (y - \gamma)x_1 - \mu x_2 \\ \dot{y} &= 1 - x_1x_2. \end{aligned} \tag{8.18}$$

As promised, the divergence of the vector field in 8.18 is $-2\mu < 0$, and so volume in the state space must contract. Moreover, there is a trapping region. To see this, first rescale 8.18 by letting

$$z = \sqrt{2}\left(y - \frac{\gamma}{2}\right).$$

Now define a function V in x_1, x_2, z space by

$$V(x_1, x_2, z) = \frac{x_1^2 + x_2^2 + z^2}{2}.$$

Then along orbits of the rescaled equations, it is easy to compute that

$$\frac{dV(x_1, x_2, z)}{dt} = x_1\dot{x}_1 + x_2\dot{x}_2 + z\dot{z} = -\mu\left(x_1^2 + x_2^2\right) + \sqrt{2}\,z,$$

which is negative along the boundary of a large enough sphere centered about the origin in x_1, x_2, z space. This means that V behaves like a Liapunov function sufficiently far from the origin, and so the orbits bend inward across the boundary of the sphere. In fact, $\dot{V} = \nabla V \cdot \mathbf{f}$ is the inner product betwen the outward normal to the sphere and the vector field $\mathbf{f}$ defined by the rescaled dynamical equations. Since this inner product is negative, the field $\mathbf{f}$ points inward and the sphere is trapping. However, z is only a rescaling of y, and so the sphere can be suitably stretched to insure the same conclusion for the original system 8.18. From our previous discussion, we can now infer that an attractor Λ exists for the orbits of the geomagnetic equations.

The equilibria in 8.18 are found by multiplying the first equation by x_2 and the second one by x_1. Setting the righthand sides of these equations to zero and then subtracting gives

$$y\left(x_2^2 - x_1^2\right) = -\gamma x_1^2.$$

Also, the third equation satisfies $x_2 = 1/x_1$ at equilibrium so that if y/μ is denoted as k^2, then it follows from $yx_2 - \mu x_1 = 0$ that $x_1^2 = k^2$ and

$$\gamma = \mu\left(k^2 - \frac{1}{k^2}\right). \tag{8.19}$$

The rest points are therefore

$$\begin{pmatrix} \bar{x}_1 \\ \bar{x}_2 \\ \bar{y} \end{pmatrix} = \begin{pmatrix} k \\ \dfrac{1}{k} \\ \mu k^2 \end{pmatrix}, \begin{pmatrix} -k \\ -\dfrac{1}{k} \\ \mu k^2 \end{pmatrix}.$$

The Jacobian of the linearized system about the first equilibrium is

$$A = \begin{pmatrix} -\mu & \mu k^2 & \dfrac{1}{k} \\ \dfrac{\mu}{k^2} & -\mu & k \\ -\dfrac{1}{k} & -k & 0 \end{pmatrix}.$$

Its eigenvalues λ_i are

$$\lambda_1 = -2\mu$$

$$\lambda_{2,3} = \pm i\sqrt{k^2 + \frac{1}{k^2}}\,.$$

It follows that the linearized system has a transient term that decays to zero as t increases (since $\lambda_1 < 0$) plus an oscillatory solution. It is therefore neutrally stable in the limit. The same result is true at the other equilibrium. Consequently, we cannot conclude anything about the behavior of the nonlinear system 8.18 by means of linearization. However, a numerical integration of these equations is quite revealing. It shows that the equilibria are actually unstable and that the orbits encircle one of them a number of times before switching suddenly to encircle the other equilibrium where it oscillates for awhile before again switching rapidly back about the original point (Figure 8.2).

The orbits are never captured by either rest point, and the limiting behavior is apparently chaotic since the number of oscillations in the vicinity of each point is unpredictable. This is reminiscent of the catastrophe model in which the current also flipped back and forth between two states in an erratic manner. Both models indicate bistable pole-reversing behavior that is aperiodic. A notable difference is that in the present model this effect is achieved without having to postulate a randomly fluctuating external magnetic field.

We cannot leave this discussion without commenting on another model whose orbits exhibit similar limiting behavior. This is the celebrated model of Lorenz whose equations are

$$\begin{aligned} \dot{x}_1 &= -\sigma x_1 + \sigma x_2 \\ \dot{x}_2 &= \quad r x_1 - x_2 - x_1 x_3 \\ \dot{x}_3 &= -b x_3 + x_1 x_2 \end{aligned} \tag{8.20}$$

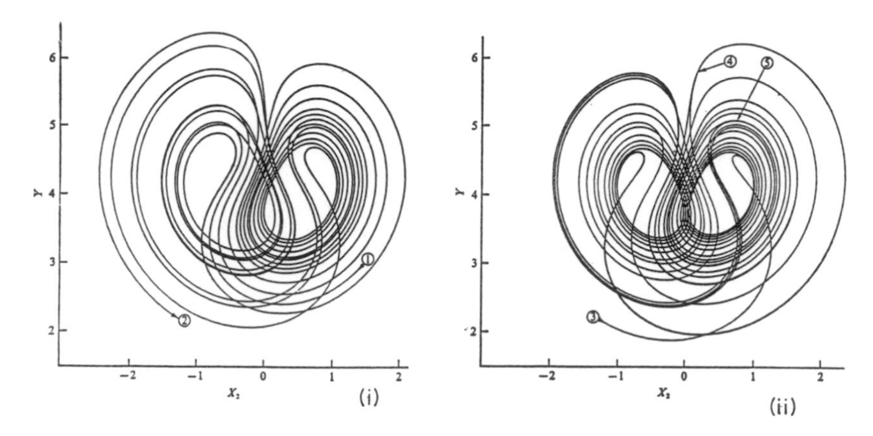

Figure 8.2 An orbit of Equations 8.18 in $\mathbf{R}^3$ projected onto the x_2, y plane, for $\mu = 1$ and $k = 2$. In (i) the orbit goes from ① to ②, and in (ii) the orbit continues until ⑤.

in which σ, r, and b are positive constants. Note, once again, that the divergence of the vector field in 8.20 is $-\sigma - 1 - b < 0$. Moreover, there is a trapping region (Exercise 8.4.7). Hence, there is a Lorenz attractor.

The equations are an approximate description of the motion of a fluid in a horizontal layer which is being heated from below. The cooler fluid at the top of the layer is denser than the warmer part at the bottom. Hence, the fluid is vertically buoyant, and so it begins to move. In fact, if the heating is rapid enough, it rolls about in alternate directions for irregular lengths of time. For certain values of r, the orbit motion is qualitatively similar to that of Figure 8.2. There are three equilibria, two of which are a.s., and one unstable as long as r is greater than one but less than some critical value $\hat{r}$. At $\hat{r}$, a Hopf bifurcation takes place, and as r continues to increase, the apparently chaotic behavior begins to be observed.

8.3 Deterministic or Random?

Recall the discussion in Section 4.2 concerning the discrete logistic equation

$$N((n+1)\,\Delta t) = N(n\,\Delta t) + rN(n\,\Delta t)\left[1 - \frac{N(n\,\Delta t)}{K}\right]\Delta t. \quad (8.21)$$

It was seen that if $r\Delta t$ is large enough, then the equilibrium solution $\overline{N} = K$ is unstable. The time has come to explain the behavior of the solution as $r\Delta t$ is allowed to increase. It is indeed more bizarre than at first appears.

To simplify matters, choose Δt to be unity and define X_n by

$$X_n = \frac{b}{a} N(n\,\Delta t),$$

where $a = 1 + r$ and $b = r/K$. Then 8.21 transforms into

$$X_{n+1} = aX_n(1 - X_n) = f(X_n). \tag{8.22}$$

Since the population has to be nonnegative, 8.22 is defined only for $0 \le X_n \le 1$.

Consider the function $f(x) = ax(1 - x)$. The equilibria of 8.21 are roots of $f(x) = x$, as one easily sees. The fixed points of f are, in fact, zero and $x = 1 - 1/a$.

The function f has a single hump in the unit interval that rises as r increases. The roots of $f(x) = x$ are simply the intersection of this hump with the 45 degree line, as shown in Figure 8.3.

It is apparent that X_n goes to zero as $n \to \infty$, if a is between zero and one. Also, X_n becomes negative, for a greater than four, which is physically untenable. Therefore, we assume that a is between one and four.

To decide whether the nonzero equilibrium $\bar{x} = 1 - 1/a$ is stable, one performs the usual linearized analysis. Let $u_n = x_n - \bar{x}$ be a "small" perturbation about $\bar{x}$. Then $x_{n+1} = f(x_n)$ is approximated by the linear

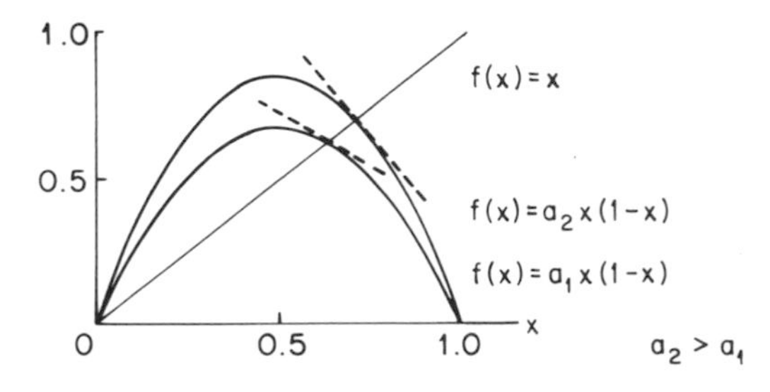

Figure 8.3 Solutions of 8.22 for different values of the constant a, seen as intersections of the straight line whose slope is unity and the humped curve.

equation

$$u_{n+1} = f'(\bar{x})u_n. \tag{8.23}$$

If u_n tends to zero as t increases, this implies that $\bar{x}$ is at least locally stable. From 8.23, it is apparent that this will be true if

$$|f'(\bar{x})| < 1.$$

That is, the origin of 8.23 is stable if the slope of f at $\bar{x}$ lies between plus and minus one. Equivalently, the angle of the slope must lie between plus and minus 45 degrees (shown by dotted lines in Figure 8.3). This slope condition translates into the statement $1 < a < 3$. When this is satisfied, all iterates of $x = f(x)$ converge to $\bar{x}$.

So far this is simply a restatement, in different notation, of what was done in Section 4.2. Less obvious is the remaining case of $a \geq 3$, which is considered next.

As a increases, the hump of the function f continues to rise. This eventually forces the tangent line at $\bar{x}$ to go below minus 45 degrees, and stability is no longer possible at this point. To see what happens, we study the second iterate of f, namely

$$f^{(2)}(x) = f(f(x)). \tag{8.24}$$

From the chain rule, we find that

$$\frac{df^{(2)}(x)}{dx} = \frac{df(f(x))}{dx}\frac{df(x)}{dx}, \tag{8.25}$$

so that if $\hat{x}_1, \hat{x}_2$ are two roots of $f(x) = \frac{1}{2}$, then since $f'(\frac{1}{2}) = 0$, it follows from 8.25 that the $\hat{x}_i$ are, in addition to $x = \frac{1}{2}$, extreme points of $f^{(2)}$. Thus, $f^{(2)}$ has three humps in the unit interval. A plot of this function is given in Figures 8.4 and 8.5.

From 8.25, it is also seen that

$$\frac{df^{(2)}(\bar{x})}{dx} = \left(\frac{df(\bar{x})}{dx}\right)^2. \tag{8.26}$$

Therefore, if the magnitude of $|f'(\bar{x})|$ is less than unity, the same must be true of $\left|\frac{df^{(2)}(\bar{x})}{dx}\right|$. But as soon as $\bar{x}$ becomes unstable at $a = 3$, then

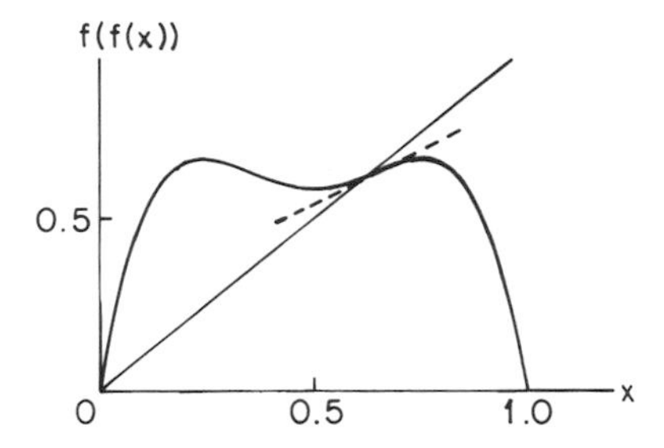

Figure 8.4 The second iterate $f(f(x))$ for the constant a equal to roughly 2.7. It has a single intersection with the 45 degree line.

$f'(\bar{x}) < -1$ and $\dfrac{df^{(2)}(x)}{dx} > 1$. Pictorially, it means that when $a > 3$ the function f forms a loop about $1 - 1/a$, which causes two new intersections with the 45 degree line to appear (Figure 8.5). The new fixed points r_1 and r_2 of $f^{(2)}$ make a sudden appearance at the very instant that the old stable point $\bar{x}$ gives way to instability. This branching of $\bar{x}$ into r_1 and r_2 at $a = 3$ is a pitchfork bifurcation not unlike those discussed in Chapter Seven.

It is clear that $f^{(2)}$ is a quartic polynomial. The four fixed points are all real only for $a > 3$ (Exercise 8.4.7). They manifest themselves as the points of intersection of $f^{(2)}$ with the 45 degree line. In Exercise 8.4.7, the fixed points r_1 and r_2 of $f^{(2)}$ map into each other under f. They constitute a period-two cycle in which the population oscillates periodically between these values over consecutive time periods.

As a continues to increase, the slope of $f^{(2)}$ at r_i eventually falls below minus one at which point a linearization argument similar to that

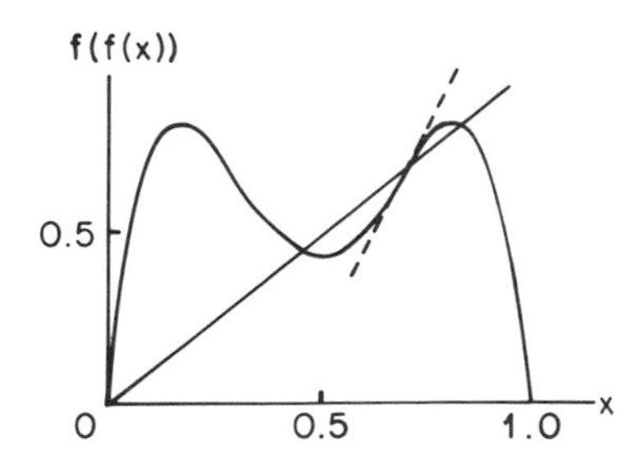

Figure 8.5 The second iterate $f(f(x))$ for the constant a equal to roughly 3.4. It has three intersections with the 45 degree line.

used previously shows that the two-cycle now becomes unstable. To examine this case further, note that the chain rule provides an extension of 8.25:

$$\frac{df^{(4)}(x)}{dx} = \frac{df^{(2)}(f^{(2)}(x))}{dx}\,\frac{df^{(2)}(x)}{dx} \tag{8.27}$$

so that, in particular,

$$\frac{df^{(4)}(r_i)}{dx} = \left(\frac{df^{(2)}(r_i)}{dx}\right)^2. \tag{8.28}$$

The fixed points of $f^{(2)}$ are clearly fixed points of $f^{(4)}$. Moreover, when the slope of $f^{(2)}$ at r_i becomes less than minus one, the slope of $f^{(4)}$ at these points will exceed plus one. This follows from 8.28. Pictorially, $f^{(4)}$ forms loops with the 45 degree line about the r_i, and as a result four new real roots $s_1 < s_2 < s_3 < s_4$ suddenly appear as intersections with this line, as illustrated in Figure 8.6 (see also Exercise 8.4.9).

Thus, once again, a pitchfork bifurcation emerges. However, in this case an unstable two-cycle gives way to a stable four-cycle. Stability of the four-cycle is respected until the slope of $f^{(4)}$ at the points s_i decreases beyond minus one when a new bifurcation takes place. This process is generic and continues to give birth to successive cycles of period 2^n as a increases. The range of a values within which each

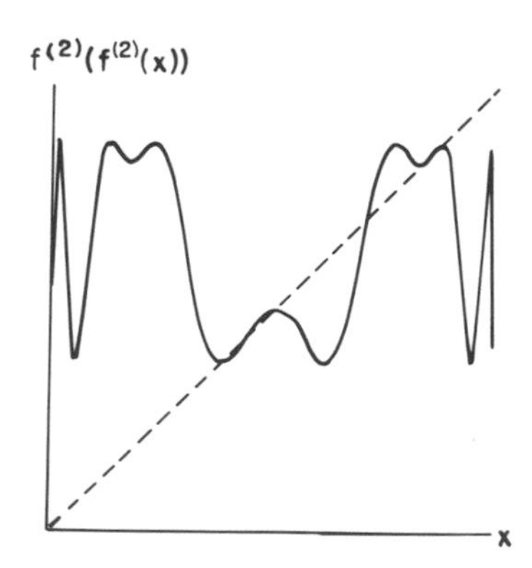

Figure 8.6 The fourth iterate $f^{(2)}(f^{(2)}(x))$ and its intersections with the 45 degree line.

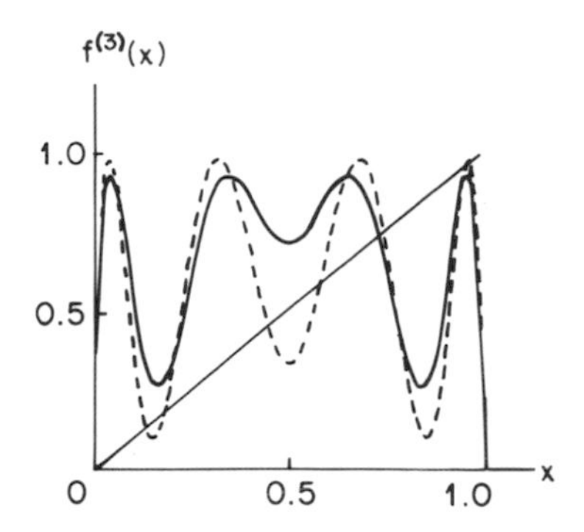

Figure 8.7 The third iterate $f^{(3)}(x)$ and its intersections with the 45 degree line.

successive cycle remains stable becomes progressively smaller. This is fairly evident if one considers that with each bifurcation more and more fixed points are crowded into the same unit interval, and so their slopes will have to shift from plus to minus one more rapidly (that is, the tangents are forced to twist about ever more quickly).

Eventually, a limiting value a is reached at which all 2^n cycles are unstable. This is known to occur at some $\hat{a}$ less than four. Beyond $\hat{a}$, a new phenomenon begins. Odd-order cycles now appear. To see how this can happen consider $f^{(3)}$. As a increases, the hills and dales of this function become more pronounced until some of the hills rise and valleys sink to touch the 45 degree line. From Figure 8.7, we see that this creates six intersections with the line. Stable and unstable period-three cycles are born.

It is important to note that the stable three-cycle does not come about from a bifurcation of previously unstable points, as has been the case up to now. Something different has happened in that truly new cycles are being formed by virtue of the up and down movement of the hills and valleys of $f^{(3)}$. As these contours intersect the 45 degree line new fixed points of $f^{(3)}$ make their appearance at the point of tangency.

It can be shown that once a period-three cycle appears there will also exist cycles of all orders. The existence of periodic cycles of arbitrarily high order results in an orbit that to all appearances is erratic in that successive iterates do not follow a simple and orderly pattern but seem to jump about. In the limiting case of $a = 4$, the orbits exhibit an essentially random motion on the unit interval, as will be shown presently. It is a remarkable fact that such unexpectedly complicated behavior should come about by iterating a benign function like $ax(1 - x)$. From a modeling standpoint, it suggests that the seemingly unpredictable fluctuations in population one sometimes observes in nature can spring from surprisingly simple causes.

We conclude by examining the case $a = 4$ more closely. To facilitate our discussion, the function f is temporarily replaced by a simple one hump function g defined by

$$g(x) = \begin{cases} 2x, & 0 \leq x < \frac{1}{2}, \\ 2(1 - x), & \frac{1}{2} \leq x \leq 1. \end{cases} \tag{8.29}$$

The connection between f and g is established later. To begin with, let us define $g^{(k)}(x)$ to be the kth iterate of x under the action of g, with $g^{(1)}$ being g itself and $g^{(0)}$ the identity map. Construct a number b in the unit interval whose binary representation $b = .b_0 b_1 b_2 \ldots$ is formed by the rule

$$b_k = \begin{cases} 0 & \text{if } g^{(k)}(x) < \frac{1}{2}, \\ 1 & \text{if } g^{(k)}(x) \geq \frac{1}{2}. \end{cases}$$

In effect, b is a coded message that tells us on which half of the unit interval successive iterates of x will fall. Now suppose that b is any number between zero and one with a binary representation $.b_0 b_1 b_2 \ldots$, and define $x = .a_0 a_1 a_2 \ldots$ by the recursion relations

$$\begin{aligned} a_0 &= b_0 \\ a_k &= (a_{k-1} + b_k) \mod 2, \qquad k = 1, 2, \ldots. \end{aligned}$$

Proceeding inductively, we see immediately that if $b_k = 0$, then

$$a_{k-1} + a_k = 0 \mod 2 \tag{8.30}$$

for $k = 1, 2, \ldots$. It is now essential to observe that $a_0 = 0$ if and only if $x < \frac{1}{2}$ and that $a_0 + a_1 = 0$ if and only if $g^{(1)}(x) < \frac{1}{2}$ (Exercise 8.4.10). By an inductive argument, one concludes in general that 8.30 is satisfied if and only if $g^{(k)}(x) < \frac{1}{2}$.

There is therefore a unique correspondence between every b and the dynamics of the orbit generated through successive iterates of x. The representation of orbit motion by means of b is an example of a technique called *symbolic dynamics*.

Since $b_k = 0$ is equivalent to $g^{(k)}(x) < \frac{1}{2}$, then for any positive integer m, the successive iterates of $\hat{x} = g^{(m)}(x)$ correspond to a number $\hat{b}$ obtained by shifting the digits in the expansion of b to the left m places and then cutting off all the digits to the left of the decimal point. Thus

$$\hat{b} = .b_m b_{m+1} \dots .$$

This establishes a correspondence between g and the shift operator:

$$\begin{array}{ccc} x & \longleftrightarrow & b \\ g^{(m)} \downarrow & & \downarrow \text{shift} \\ \hat{x} & \longleftrightarrow & \hat{b} \end{array}$$

An immediate consequence of this fact is that x is periodic with period k, namely $g^{(k)}(x) = x$, if and only if the corresponding b satisfies

$$b = \underbrace{.b_0 b_1 \dots b_{k-1}}\,\underbrace{b_0 b_1 \dots b_{k-1}} \dots .$$

There are 2^k ways of arranging a string of k zero and one digits. Hence, there are exactly 2^k cycles of period k under the mapping g. It follows that there can be only a countable number of periodic cycles, and so there are necessarily an uncountable number of aperiodic ones!

Consider the implications of this. It is known from probability theory that except for a set of initial b of measure zero every binary expansion $\{b_k\}$ of b can be realized as a succession of tosses of a fair coin in which heads are indicated by a zero and tails by unity. Thus, $\{b_k\}$ is effectively indistinguishable from a random process in which successive coin tosses give equally likely results. This means that the aperiodic orbits jump about from one half of the unit interval to the other in an apparently random way. If each interval half is defined as a state of x, we see that an orbit of x is bistable and the amount of time it occupies in each state before flipping to the other is random, just as in the coin tossing experiment. This is reminiscent of the geomagnetic model (and, for that matter, the Lorenz model) in which orbits cycle about an equilibrium for an apparently random length of time before flipping over to cycle about the other equilibrium only to be eventually captured again in a neighborhood of the first equilibrium, and so on. Such behavior, appropriately enough, is called chaotic.

We now return to the function $f(x) = 4x(1 - x)$. Let

$$x = h(\theta) = \sin^2 \frac{\pi\theta}{2}.$$

It is easy to see that h defines a continuous and invertible map between x and θ on the unit interval. Now observe that

$$4x(1 - x) = 4\sin^2 \frac{\pi\theta}{2}\left(1 - \sin^2 \frac{\pi\theta}{2}\right) = \left(2\sin \frac{\pi\theta}{2} \cos \frac{\pi\theta}{2}\right)^2 = \sin^2 \pi\theta.$$

From this, we see that if $\hat{x} = f(x)$ and if $\hat{\theta}$ is defined by $\hat{x} = h(\hat{\theta})$, then

$$\sin \frac{\pi\hat{\theta}}{2} = \sin \pi\theta.$$

This relation has two roots $\hat{\theta} = 2\theta$ and $\hat{\theta} = 2(1 - \theta)$. This shows that $\hat{\theta}$ defines a function $g(\theta)$ by

$$\hat{\theta} = g(\theta) = \begin{cases} 2\theta, & \theta < \frac{1}{2}, \\ 2(1 - \theta), & \theta \geq \frac{1}{2}. \end{cases}$$

Thus, h maps the "tent function" g into f and conversely. Specifically, one has $g = h^{-1}fh$. For any nonnegative integer k, it follows immediately that $g^{(k)} = h^{-1}f^{(k)}h$ in which $f^{(k)}$ is the kth iterate of f. Consequently, periodic and aperiodic orbits under the action of g map into orbits having the same properties with respect to f. Letting pr indicate "probability of" one also finds that

$$\begin{aligned} &\mathrm{pr}\left(f^{(k)}(x) < \tfrac{1}{2}\right) = \mathrm{pr}\left(f^{(k)}h(\theta) < \tfrac{1}{2}\right) = \mathrm{pr}\left(hg^{(k)}(\theta) < \tfrac{1}{2}\right) \\ &= \mathrm{pr}\left(g^{(k)}(\theta) < \frac{2}{\pi} \arcsin \frac{1}{\sqrt{2}}\right) = \mathrm{pr}\left(g^{(k)}(\theta) < \tfrac{1}{2}\right) = \mathrm{pr}(b_k = 0) = \tfrac{1}{2}. \end{aligned}$$

Therefore, the map $f^{(k)}$ exhibits the same random behavior as $g^{(k)}$. This completes our discussion of the logistic map.

8.4 Exercises

8.4.1 Write 8.5 as $\dot{\mathbf{N}} = \mathbf{f}(\mathbf{N})$ and show that $\alpha + \beta > 2$ implies that $\text{Div}\,\mathbf{f} < 0$ asymptotically as t increases.

8.4.2 Examine the borderline case of $\alpha + \beta = 2$ in 8.5, and discuss the kind of orbits one obtains in this case. This can be done explicitly. Hint: Consider 8.12 and 8.13. What is the attractor?

8.4.3 Carry out a numerical integration of 8.12 in order to verify the limiting properties of its orbits as mentioned in the text.

8.4.4 Reconsider Equation 6.5.10 in which strongly competing species can coexist only if both are victims of a third predator species. Do a numerical study to show that as the predation rates increase the orbits may begin to exhibit chaotic behavior.

8.4.5 Let $k = 1$ in Equation 8.18. Compute the form of the attractor and show that no field reversals are possible in the limit. Sketch the orbits and interpret the results.

8.4.6 Let $r < 1$ in the Lorenz equations 8.20. Show that there is a single rest state in this case that is a.s.

If $r > 1$, there are three equilibria. One is at the origin, which is now unstable, and two others are stable points P_1 and P_2 located symmetrically about the x_3 axis. Show this. Prove that if $\sigma - b - 1 < 0$, then P_1 and P_2 remain stable for all $r > 1$. Otherwise, there is an $\hat{r}$ at which a Hopf bifurcation takes place and beyond which P_1 and P_2 are unstable.

8.4.7 Show that a sufficiently large sphere about the origin is a trapping region for 8.20. Use an argument similar to that given in the text for system 8.18.
Hint: Rescale the variable x_3 by adding a suitable constant.

8.4.8 If $f(x) = ax(1 - x), 0 < x < 1$, show that $f^{(2)}(x) = f(f(x))$ is a quartic polynomial whose roots are zero and $1 - 1/a$ in addition to two roots r_1 and r_2 that are complex for $a < 3$ and real for $a \geq 3$. Verify that $r_2 = f(r_1)$ and $r_1 = f(r_2)$.

8.4.9 Continuing with the previous exercise, consider $f^{(4)}(x) = f^{(2)}(f^{(2)}(x))$. Using a calculator or computer, find the roots. In addition to the four roots of $f^{(2)}$ found previously, there are four new real roots when a is large enough. To be specific, let $a = 3.5$. Show that these new roots $s_1 < s_2 < s_3 < s_4$ constitute a period-four cycle under the action of f. In fact, $s_1 \to s_3 \to s_2 \to s_4 \to s_1$.

8.4.10 Establish that relation 8.30 is equivalent to the condition $g^{(k)}(x) < \frac{1}{2}$.

8.4.11 The negative divergence condition of Section 8.1 has an interesting consequence when applied to a region B in the plane. Suppose that B is some large circle about the origin in $\mathbf{R}^2$. Then if $\operatorname{Div} \mathbf{f}$ is a negative constant this implies that B contains no cycles of system 8.1. Show this by an application of Green's Theorem in calculus.

8.4.12 Establish that the damped pendulum, whose equation of motion is

$$\ddot{\theta} + r\dot{\theta} + \frac{g}{l} \sin \theta = 0,$$

has no cycles. Apply the previous exercise.

Chapter Nine

There Is a Better Way

9.1 Conditions Necessary for Optimality

Dynamic models can be useful in predicting the long-term behavior of physical processes especially when it is not possible to penetrate into the actual system itself. We hope that the preceding chapters have been at least partially convincing about this. There is however another context for dynamic modeling in terms of being able to optimize the behavior of a given system. We have seen how feedback controls can modify performance. In this chapter, we wish to show that with a proper choice of control one can not only modify but also improve performance of a system. This entails an introduction to certain aspects of optimal control theory. This topic is not only of interest in itself involving, as it does, linear and nonlinear differential equations (the core tool of this book), but also because it complements some of the material discussed earlier.

In particular, we return to the problem of the inverted pendulum, which is unstable in its upright position. We were able to stabilize it by a suitable linear feedback control in Chapter Three. Later, it will be seen that this control is the consequence of a natural optimality condition.

The fish harvesting model of Chapter Six is also re-examined by asking for the harvesting policy that maximizes the net profit from fishing. Finally, we come full circle and return once again to the harmonic oscillator, the model with which this book began. We now ask for a damping force that will bring the spring back to its position of rest in minimum time. This problem, like the others discussed in this chapter, has important practical implications.

It is necessary to begin by stating a set of conditions that must be met in order to insure optimality. These are stated without proof since a satisfactory discussion would burden us with considerable detail. Several lucid treatments do exist, however, and it is to these sources that the reader is directed. Moreover, in keeping with the character of this book, we are more interested in the implication of optimality conditions to modeling than in a complete derivation.

In Chapter Three, we encountered feedback control systems of the form

$$\begin{aligned}\dot{x}_1 &= x_2\\ \dot{x}_2 &= a_{21}x_1 + a_{22}x_2 + u\end{aligned}$$

in which $u(t)$ is a control function. More generally, one may wish to consider a first-order system of the form

$$\dot{\mathbf{x}} = \mathbf{f}(\mathbf{x}, \mathbf{u}) \tag{9.1}$$

in which $\mathbf{x}$ is a vector of $\mathbf{R}^k$, $\mathbf{u}$ a vector in $\mathbf{R}^m$, and $\mathbf{f}$ is a smooth vector-valued function of $\mathbf{x}$ and $\mathbf{u}$ with components in $\mathbf{R}^m$. In all our applications $k \leq 2$ and $m = 1$. For any initial value $\mathbf{x}(0) = \mathbf{x}_0$, we assume that there is a unique smooth orbit of 9.1 for each choice of function $\mathbf{u}$ in a certain admissible class $\mathscr{F}$. Each choice of an admissible $\mathbf{u}$ alters the right side of 9.1, and this induces a different solution to emerge from $\mathbf{x}_0$. For this reason, $\mathbf{u}$ is called a *control function*.

It is important for applications that $\mathbf{u}$ be allowed to have a finite number of discontinuities on any given time interval $0 \leq t \leq T$, and so $\mathscr{F}$ consists of piecewise continuous functions on the real axis that are restricted to take on values in some closed set U in $\mathbf{R}^m$ (since $m = 1$ in our examples, U is a closed subset of the real axis).

Now suppose that our goal is to minimize the integral

$$\int_0^T f_0(\mathbf{x}(t), \mathbf{u}(t))\, dt \tag{9.2}$$

over some interval $[0, T]$ by a suitable choice of admissible $\mathbf{u}$. f_0 is a given smooth scalar-valued function of $\mathbf{x}$ and $\mathbf{u}$, specific examples of which will be given below.

Each $\mathbf{u}$ in $\mathscr{F}$ determines a solution of 9.1 for each given $\mathbf{x}_0$ and therefore a corresponding value of the integral 9.2. Let $\hat{\mathbf{u}}$ be a control in $\mathscr{F}$ that minimizes the integral and denote the resulting orbit by $\hat{\mathbf{x}}$. Define a function H by

$$H(\mathbf{x}, \boldsymbol{\lambda}, \mathbf{u}) = \lambda_0 f_0(\mathbf{x}, \mathbf{u}) + \sum_{j=1}^{k} \lambda_j f_j(\mathbf{x}, \mathbf{u}), \tag{9.3}$$

where λ_0 is some constant and λ_i are solutions to the following linear differential equations:

$$\dot{\lambda}_i = -\lambda_0 \frac{\partial f_0(\mathbf{x}, \mathbf{u})}{\partial x_i} - \sum_{j=1}^{k} \lambda_j \frac{\partial f_j(\mathbf{x}, \mathbf{u})}{\partial x_i}, \qquad 1 \leq i \leq k. \tag{9.4}$$

In vector notation, 9.4 becomes

$$\dot{\boldsymbol{\lambda}} = -\lambda_0 f_{\mathbf{x}}(t) - A(t)\boldsymbol{\lambda} \tag{9.5}$$

in which

$$\boldsymbol{\lambda} = \begin{pmatrix} \lambda_1 \\ \vdots \\ \lambda_k \end{pmatrix}, \qquad f_{\mathbf{x}} = \begin{pmatrix} \dfrac{\partial f_0}{\partial x_1}(\mathbf{x}, \mathbf{u}) \\ \vdots \\ \dfrac{\partial f_0}{\partial x_k}(\mathbf{x}, \mathbf{u}) \end{pmatrix},$$

and $A(t)$ is the transpose of the Jacobian matrix of $\mathbf{f}$ with respect to $\mathbf{x}$, evaluated at $\mathbf{x}(t), \mathbf{u}(t)$:

$$A(t) = \begin{pmatrix} \dfrac{\partial f_1(\mathbf{x}(t), \mathbf{u}(t))}{\partial x_1} & \cdots & \dfrac{\partial f_k(\mathbf{x}(t), \mathbf{u}(t))}{\partial x_1} \\ \vdots & & \vdots \\ \dfrac{\partial f_1(\mathbf{x}(t), \mathbf{u}(t))}{\partial x_k} & \cdots & \dfrac{\partial f_k(\mathbf{x}(t), \mathbf{u}(t))}{\partial x_k} \end{pmatrix}. \tag{9.6}$$

Note that one may alternately express Equations 9.1 and 9.5 as

$$\dot{x}_i = \frac{\partial H}{\partial \lambda_i}$$
$$\dot{\lambda}_i = -\frac{\partial H}{\partial x_i} \tag{9.7}$$

for $1 \leq i \leq k$.

We can now state the following proposition.

Theorem 9.1 *Let* $\hat{\mathbf{u}}$ *be a control in* $\mathcal{F}$ *which minimizes the integral* 9.2, *and let* $\hat{\mathbf{x}}$ *and* $\hat{\boldsymbol{\lambda}}$ *be the solutions to* 9.1 *and* 9.5 (*or, equivalently, of* 9.7) *corresponding to* $\hat{\mathbf{u}}$. *Then for all* $\mathbf{v}$ *in* U, *it is necessarily true that*

$$H(\hat{\mathbf{x}}(t), \hat{\boldsymbol{\lambda}}(t), \mathbf{v}) \leq H(\hat{\mathbf{x}}(t), \hat{\boldsymbol{\lambda}}(t), \hat{\mathbf{u}}(t)) \tag{9.8}$$

in which λ_0 *has the value minus one and* $\hat{\boldsymbol{\lambda}}(T) = \mathbf{0}$.

Now suppose that the orbit that begins at $\mathbf{x}_0$ is further constrained to pass through a given terminal value $\mathbf{x}_T$ at time T. The class $\mathcal{F}$ is redefined in this case to include all admissible controls that transfer the state $\mathbf{x}$ from $\mathbf{x}_0$ to $\mathbf{x}_T$. We assume that $\mathcal{F}$ is nonempty and that a unique solution to 9.1 exists that satisfies the additional constraint. Then 9.8 is still true except that it can now happen that λ_0 is zero. When it turns out that λ_0 is indeed zero, then $\hat{\boldsymbol{\lambda}}(t)$ is nonzero for all $0 \leq t \leq T$. Moreover, the final value of $\hat{\boldsymbol{\lambda}}$ is unrestricted whenever $\mathbf{x}$ satisfies a terminal constraint.

Finally, if T itself is not specified but is allowed to vary, then it is additionally true that

$$H(\hat{\mathbf{x}}(t), \hat{\boldsymbol{\lambda}}(t), \hat{\mathbf{u}}(t)) = 0 \tag{9.9}$$

for all $0 \leq t \leq T$.

Theorem 9.1 is a special case of what is known as the *maximum principle*. Although the statement can be extended to allow, for example, orbits to terminate on some implicitly defined k-dimensional surface rather than at a point of $\mathbf{R}^k$, it suffices for the applications we have in mind. What remains to be done, then, is to bring the theorem to life by considering specific cases.

Example 9.1 In Section 3.3, we saw that the inverted pendulum had an unstable equilibrium that could be stabilized in its upright position by imposing a control in the form of a lateral movement which was linear in angle θ and velocity $\dot{\theta}$. An alternate point of view is to find a control u that minimizes the deviation of θ and $\dot{\theta}$ from zero while restricting the total use of u itself. As will be seen shortly, the optimal u again turns out to be linear in θ and $\dot{\theta}$.

Letting x_1 and x_2 represent θ and $\dot{\theta}$, the pendulum equations linearized about the unstable point zero in $\mathbf{R}^2$ are given by equations 3.28. They are rewritten here after first denoting g/l by a and $-v/ml$ as u:

$$\begin{aligned} \dot{x}_1 &= x_2 \\ \dot{x}_2 &= ax_1 - rx_2 + u. \end{aligned} \tag{9.10}$$

In matrix notation 9.10 becomes

$$\dot{\mathbf{x}} = A\mathbf{x} + u\mathbf{b}, \tag{9.11}$$

where

$$A = \begin{pmatrix} 0 & 1 \\ a & -r \end{pmatrix} \quad \text{and} \quad \mathbf{b} = \begin{pmatrix} 0 \\ 1 \end{pmatrix}.$$

Since we wish the solution to 9.11 to deviate as little as possible from zero while insuring that the restoring force is also as small as possible, we choose to minimize the integral

$$\frac{1}{2}\int_0^T \left(x_1^2 + x_2^2 + u^2\right) dt = \int_0^T f_0(\mathbf{x}, \mathbf{u})\, dt$$

over some time interval $[0, T]$. Later T will be allowed to go to infinity.

Theorem 9.1 permits us to write down a set of necessary conditions for optimality. To do this, let us first define a function H according to 9.3:

$$H(\mathbf{x}, \boldsymbol{\lambda}, u) = \tfrac{1}{2}\lambda_0\left(x_1^2 + x_2^2 + u^2\right) + \lambda_1 x_2 + \lambda_2\left(ax_1 - rx_2 + u\right). \tag{9.12}$$

Since $\mathbf{x}(T)$ is unrestricted, λ_0 can be taken to be -1. The scalar components of Equation 9.5 become

$$\begin{aligned} \dot{\lambda}_1 &= x_1 - a\lambda_2 \\ \dot{\lambda}_2 &= x_2 - \lambda_1 + r\lambda_2, \end{aligned} \tag{9.13}$$

or

$$\dot{\boldsymbol{\lambda}} = \mathbf{x} - A^{\mathrm{T}}\boldsymbol{\lambda}. \tag{9.14}$$

In this example, u is unrestricted in the sense that it can take on any value in the set $U = \mathbf{R}^1$. Hence, the maximum condition 9.8 can be rephrased as

$$\frac{\partial H}{\partial v}\left(\hat{\mathbf{x}}(t), \hat{\boldsymbol{\lambda}}(t), v\right) = 0 \tag{9.15}$$

for all $0 \le t \le T$. Therefore, when $v = \hat{u}(t)$, one has

$$-\hat{u} + \hat{\lambda}_2 = 0.$$

It follows that the optimal u satisfies

$$\hat{u} = \mathbf{b}^{\mathrm{T}}\hat{\boldsymbol{\lambda}}. \tag{9.16}$$

We claim that the $\hat{\boldsymbol{\lambda}}$ that corresponds to $\hat{u}$ is given by $\hat{\boldsymbol{\lambda}} = -K\hat{\mathbf{x}}$, where $K(t)$ is a suitable 2 by 2 matrix whose components are smooth in t. To verify this, substitute $\hat{\boldsymbol{\lambda}} = -K\hat{\mathbf{x}}$ and $\hat{u} = \mathbf{b}^{\mathrm{T}}\hat{\boldsymbol{\lambda}}$ into the system of four first-order equations

$$\begin{aligned} \dot{\mathbf{x}} &= A\mathbf{x} + u\mathbf{b} \\ \dot{\boldsymbol{\lambda}} &= \mathbf{x} - A^{\mathrm{T}}\boldsymbol{\lambda}. \end{aligned} \tag{9.17}$$

We obtain

$$\begin{aligned} \frac{d\hat{\mathbf{x}}}{dt} &= (A + \mathbf{b}\mathbf{b}^{\mathrm{T}}K)\hat{\mathbf{x}} \\ -\frac{dK}{dt}\hat{\mathbf{x}} - K\frac{d\hat{\mathbf{x}}}{dt} &= (I + A^{\mathrm{T}}K)\hat{\mathbf{x}}. \end{aligned}$$

Multiply the first of these relations by K and then add both equations to obtain

$$(\dot{K} + KA + A^{\mathrm{T}}K + I - K\mathbf{b}\mathbf{b}^{\mathrm{T}}K)\hat{\mathbf{x}} = 0,$$

which is satisfied for all possible $\hat{\mathbf{x}}$ in $\mathbf{R}^2$ provided K satisfies the matrix differential equation

$$-\dot{K} = KA + A^{\mathrm{T}}K + I - K\mathbf{b}\mathbf{b}^{\mathrm{T}}K. \tag{9.18}$$

This is a matrix version of a scalar nonlinear differential equation known as the *Riccati equation*.

Since $\mathbf{x}(T)$ is unrestricted we know from Theorem 9.1 that $\hat{\boldsymbol{\lambda}}(T) = 0$. Therefore $K(T) = 0$. Now let $T \to \infty$. The terminal condition on K then becomes irrelevant, and it can be shown by direct computation that a symmetric positive-definite constant matrix K satisfies 9.18 for all r small enough, and therefore also 9.17 (Exercise 9.4.1). With K constant, one has $\dot{K} = 0$, and so 9.18 reduces to

$$KA + A^{\mathrm{T}}K + I - K\mathbf{b}\mathbf{b}^{\mathrm{T}}K = 0.$$

This consists of four algebraic equations, one for each component of $K = \begin{pmatrix} k_1 & k_1 \\ k_2 & k_3 \end{pmatrix}$. The optimal u is therefore obtained from 9.16 in terms of K as

$$\hat{u}(t) = \mathbf{b}^{\mathrm{T}}\hat{\boldsymbol{\lambda}}(t) = -\mathbf{b}^{\mathrm{T}}K\hat{\mathbf{x}}(t). \tag{9.19}$$

We now make the trivial but important observation that every $\mathbf{x}$ lies on an optimal trajectory, namely the unique trajectory that begins at $\mathbf{x}$. Therefore, any $\mathbf{x}$ in the state space lies on an optimal path. It follows that the optimal control can be synthesized explicitly as a function of $\mathbf{x}$ by writing $\hat{u} = -\mathbf{b}^{\mathrm{T}}K\mathbf{x}$, for all $\mathbf{x}$.

Since $\mathbf{b} = \begin{pmatrix} 0 \\ 1 \end{pmatrix}$, this means that $\hat{u} = k_2x_1 + k_3x_2$, which is a feedback control, linear in x_1 and x_2. In Section 3.3, we needed to assume that u (namely $v = -mlu$) was linear in x_1 and x_2, whereas now it is a consequence of optimality.

There is one important detail to clear up however. Because of 9.19, the dynamical equation $\dot{\mathbf{x}} = A\mathbf{x} + u\mathbf{b}$ becomes

$$\dot{\mathbf{x}} = (A - bb^{\mathrm{T}}K)\mathbf{x}. \tag{9.20}$$

It was shown earlier that the upright equilibrium of the pendulum is asymptotically stable if the lateral movement is properly chosen. Does our optimal u also have the property of making the zero rest state of

Equation 9.20 a.s.? The answer is yes. To see this, let $V(\mathbf{x}) = \mathbf{x}^{\mathrm{T}}K\mathbf{x}$, which is positive, except for a zero at $\mathbf{x} = \mathbf{0}$, since K is positive definite. Along every orbit of 9.20, we have, by direct computation,

$$\dot{V} = (K\mathbf{x})^{\mathrm{T}}\mathbf{x} + (\dot{\mathbf{x}})^{\mathrm{T}}K\mathbf{x} = \mathbf{x}^{\mathrm{T}}(KA + AK - 2K\mathbf{b}\mathbf{b}^{\mathrm{T}}K)\mathbf{x},$$

which by 9.10 is

$$\dot{V} = -\mathbf{x}^{\mathrm{T}}(K\mathbf{b}\mathbf{b}^{\mathrm{T}}K + I)\mathbf{x}.$$

Since K is positive definite, the righthand side is negative except at the origin. Thus, V is a Liapunov function and the origin is globally a.s. We conclude that the optimal control belongs to the class of linear feedback u that make $\mathbf{x}(t)$ tend to the equilibrium state of zero as t goes to infinity.

9.2 Fish Harvesting

We now return to the exploitation of a fishery that was previously discussed in Example 6.7. Assuming logistic growth, the dynamics of the fishery is captured by

$$\dot{x} = rx\left(1 - \frac{x}{K}\right) - \nu Ex \tag{9.21}$$

in which ν is a catchability coefficient and E is the harvesting effort, which can vary over time. The net income from the catch is

$$R = E(\nu px - c), \tag{9.22}$$

where p is the per-unit price of the catch and c is the cost per-unit effort. If E increases with positive income and decreases otherwise, the dynamics of E may be written as

$$\dot{E} = \alpha E(\nu px - c). \tag{9.23}$$

The system consisting of 9.22 and 9.23 has an a.s. equilibrium at $\bar{x} = c/p$, $\bar{E} = f(\bar{x})/\nu\bar{x}$, where the net income is zero (Exercise 6.5.6). This corresponds to what would be expected in a common access fishery in which each fisherman continues to harvest as long as the net income is positive. Not to do so would mean that whatever profits from a haul that could be realized are being relinquished to the other fishermen. Since investments in fishing gear and boats is largely irreversible, there is every incentive to continue fishing until R has been driven to zero. This is an example of what has been called "the tragedy of the commons".

But now suppose that there is a sole owner of the fishery who decides to maximize the net income over some arbitrary time horizon T. Instead of trying to bring R down to zero, it may be wiser to restrain the fishing effort so that a positive net income can accrue. In effect, this means that instead of discounting the future, one recognizes that a future gain can have some value in the present. Accordingly the following expression, which depends on the harvest rate E, represents the total income over the given time period:

$$\int_0^T e^{-\delta t}(\nu p x - c) E\, dt.$$

The quantity $\delta \geq 0$ is a discount factor.

From now on, we denote E by u. The problem is to choose u so that the integral

$$\int_0^T e^{-\delta t}(\nu p x - c) u\, dt = \int_0^T f_0(x, u)\, dt \tag{9.24}$$

is maximized. We will actually deal with an equivalent problem of minimizing the integral obtained by replacing the integrand with its negative. Total effort u cannot be unbounded since fleet size and the investment in gear is limited. This translates into an inequality constraint

$$0 \leq u \leq u_{\mathrm{m}} \tag{9.25}$$

for some maximum effort u_{m}.

Let us fix the final value of the fish stock to be a nonnegative quantity. Specifically $x(T)$ takes on a value $x_T \geq 0$. Equation 9.21 then becomes

$$\dot{x} = f(x) - \nu u x, \tag{9.26}$$

where $f(x)$ is of logistic form and $x(0) = x_0$ and $x(T) = x_T$ are given. In order to employ Theorem 9.1, we define the expression

$$H(x, \lambda, u) = -\lambda_0 e^{-\delta t}(\nu p x - c)u + \lambda(f(x) - \nu u x). \tag{9.27}$$

The quantity λ_0 cannot be zero for otherwise we reach a contradiction (Exercise 9.4.2). Hence, $\lambda_0 = -1$, and 9.27 is rewritten as

$$H(x, \lambda, u) = \left[e^{-\delta t}(\nu p x - c) - \lambda \nu x\right] u + \lambda f(x). \tag{9.28}$$

Denote the term within braces by $\sigma(t)$. For each $0 \leq t \leq T$, we know from 9.8 that

$$H(\hat{x}(t), \hat{\lambda}(t), v) \leq H(\hat{x}(t), \hat{\lambda}(t), \hat{u}(t))$$

for all $0 \leq v \leq u_{\mathrm{m}}$. It follows that $\hat{u} = u_{\mathrm{m}}$ if $\sigma(t) > 0$, while $\hat{u} = 0$ if $\sigma(t) < 0$. What about $\sigma(t) = 0$?

Letting σ be zero in 9.28 gives

$$\lambda = e^{-\delta t}\left(p - \frac{c}{\nu x}\right), \tag{9.29}$$

and so

$$\dot{\lambda} = e^{-\delta t}\left[\frac{c\dot{x}}{\nu x^2} - \delta\left(p - \frac{c}{\nu x}\right)\right]. \tag{9.30}$$

However, we also know that

$$\dot{\lambda} = -\frac{\partial H}{\partial x} = -e^{-\delta t}\nu p u - \lambda(f'(x) - \nu u),$$

which becomes, by virtue of 9.29,

$$\dot{\lambda} = -e^{-\delta t}\left[\nu p u + \left(p - \frac{c}{\nu x}\right)(f'(x) - \nu u)\right]. \tag{9.31}$$

Since $f(x) - \nu u x = \dot{x}$ and $f'(x) = f(x)/x - rx/K$, 9.31 simplifies to

$$\dot{\lambda} = -e^{-\delta t}\left(pf'(x) - \frac{c\dot{x}}{\nu x^2} + \frac{cr}{\nu K}\right). \tag{9.32}$$

Equating 9.30 and 9.32 results in the expression

$$pf'(x) + \frac{cr}{\nu K} = \delta\left(p - \frac{c}{\nu x}\right). \tag{9.33}$$

This is a quadratic equation in x having two real roots, one of which is positive (Exercise 9.4.3). Call it $\tilde{x}$.

The orbit takes on the unique constant value $\tilde{x}$ on any time interval I on which $\sigma(t)$ is zero. It follows that $\dot{x} = 0$ on I and the optimal u becomes $\hat{u} = f(\tilde{x})/\nu\tilde{x}$. Except when $\dot{x}$ is zero (where σ is zero), the orbit either increases or decreases as rapidly as possible since u is at one of its extreme values of zero or u_{m} (assume that u_{m} is large enough so that $\dot{x} < 0$ when $\hat{u} = u_{\mathrm{m}}$). Therefore, a given final value x_T is generally unreachable unless the orbit is able to switch its direction at some intermediate time. This can only happen if σ passes through zero along the way.

Now suppose that $x(\tilde{t}) > \tilde{x}$ for some $\tilde{t}$. Any subinterval I on which $\sigma = 0$ necessarily excludes $\tilde{t}$. We claim, in fact, that $\sigma(\tilde{t})$ is positive. If $\sigma(\tilde{t}) < 0$, this implies that $\hat{u}(\tilde{t}) = 0$, and so $\dot{x} > 0$. Thus, the orbit always increases and x_T can never be attained. A similar argument shows that if $x(\tilde{t}) < \tilde{x}$ from some $\tilde{t}$, then $\sigma(\tilde{t})$ is negative. We conclude that the optimal u is determined by

$$\hat{u}(t) = \begin{cases} 0 & \text{if } \hat{x}(t) < \tilde{x} \\ u_{\mathrm{m}} & \text{if } \hat{x}(t) > \tilde{x} \\ \dfrac{f(\tilde{x})}{\nu\tilde{x}} & \text{if } \hat{x}(t) = \tilde{x}. \end{cases} \tag{9.34}$$

Observe that $\hat{u}$ is a feedback control since it depends explicitly on the state $\hat{x}$ for all t.

Once the value $\tilde{x}$ is attained the orbit cannot change direction until it is forced to do so in order to satisfy the final value x_T. This can be rephrased more quaintly by thinking of $\hat{x}(t)$ as the path taken by a traveler who finds it expedient to move as quickly as possible to the "turnpike" $x = \tilde{x}$ where he remains for as long as possible until obliged to veer off to reach the final destination x_T. The optimal path defines what is not inappropriately called a *linear turnpike*.

Let

$$\rho(x) = \left(p - \frac{c}{\nu x}\right) f(x).$$

Then 9.33 is equivalent to

$$\frac{d\rho}{dx} = \delta\left(p - \frac{c}{\nu x}\right). \tag{9.35}$$

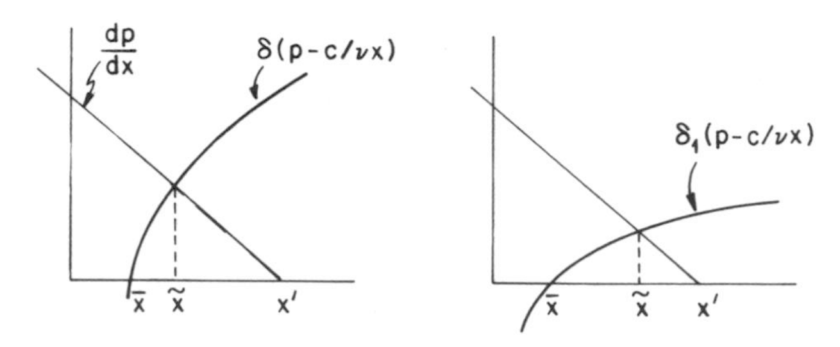

Figure 9.1 Graph of the relation 9.35 for $0 < \delta_1 < \delta$.

The left side of this relation is linear in x, and it intersects the curve defined by the right side at $\tilde{x}$ (Figure 9.1). Observe that if $\delta = 0$ then 9.35 forces $\tilde{x}$ to take on the value x' shown in the figure. On the other hand, $\tilde{x}$ tends to $\bar{x} = c/\nu p$ as $\delta \to \infty$. A glance at 9.35 shows, in fact, that $p - c/\nu x$ is required to go to zero as δ increases since $d\rho/dx$ is bounded for all $x \geq 0$. Thus, the net revenue $(p\nu x - c)u$ is zero when the future is totally discounted ($\delta = \infty$). This corresponds to what happens in an open access fishery. In general, however, $\tilde{x} > \bar{x}$ and the net revenue is positive. One sees from this that an optimal harvesting policy $\hat{u}$ prevents revenue from being dissipated through economic and biological overfishing.

Denote the total catch rate $\nu u(t)x(t)$ by $h(t)$ and in the interest of simplicity lets ignore harvesting costs from now on. Then the revenue R is a linear function of h: $R(h) = ph$. A more general situation is one in which the revenue is some nonlinear function $R(h) = p(h)h$. We assume that $R'(h) > 0$, meaning that revenue increases with catch. Let us also suppose that revenue increases occur at a decreasing rate to reflect the notion that as supply gets larger the price drops: $R''(h) < 0$. Our objective in this case is to maximize the discounted net revenue as given by

$$\int_0^T e^{-\delta t} R(h(t))\, dt = \int_0^T f_0(h)\, dt.$$

The function h plays the role of a control function, and we require that $h \geq 0$. The dynamics of fish growth is expressed in terms of h as

$$\dot{x} = f(x) - h(t)$$

with $x(0) = x_0$ and $x(T) = x_T$ given. As before, replace f by its negative thereby reducing the problem to one of minimization. Next, form a function

$$H(x, \lambda, h) = -\lambda_0 e^{-\delta t} R(h) + \lambda(f(x) - h).$$

For the same reason as before, λ_0 cannot be zero, and so our problem is one of maximizing the expression

$$H(\hat{x}, \hat{\lambda}, v) = \left(e^{-\delta t} R(v) - \lambda v\right) + \hat{\lambda} f(\hat{x}) \tag{9.36}$$

over all $v \geq 0$.

We now suppose that an optimal harvesting policy leads to a $h(t)$ that is never zero. This is reasonable in view of the fact that harvesting costs are ignored. Thus, the maximization of 9.36 takes place over the open set $v > 0$, and so at the optimum one must necessarily have

$$\frac{\partial H}{\partial v} = \left(e^{-\delta t} R'(v) - \hat{\lambda}\right) = 0.$$

That is, $\hat{\lambda}$ satisfies the relation

$$\lambda(t) = e^{-\delta t} R'\left(\hat{h}(t)\right). \tag{9.37}$$

Hence,

$$\dot{\lambda} = -\delta e^{-\delta t} R'(\hat{h}) + e^{-\delta t} R''(\hat{h}) \frac{d\hat{h}}{dt}. \tag{9.38}$$

From 9.37, it is also true that

$$\dot{\lambda} = -\frac{\partial H}{\partial x} = -\lambda f'(x) = -e^{-\delta t} R'(h) f'(x). \tag{9.39}$$

Equating 9.38 and 9.39 gives

$$\frac{d\hat{h}}{dt} = \frac{[\delta - f'(\hat{x})] R'(\hat{h})}{R''(\hat{h})}. \tag{9.40}$$

By taking 9.40 together with the equation

$$\frac{d\hat{x}}{dt} = f(\hat{x}) - \hat{h}, \tag{9.41}$$

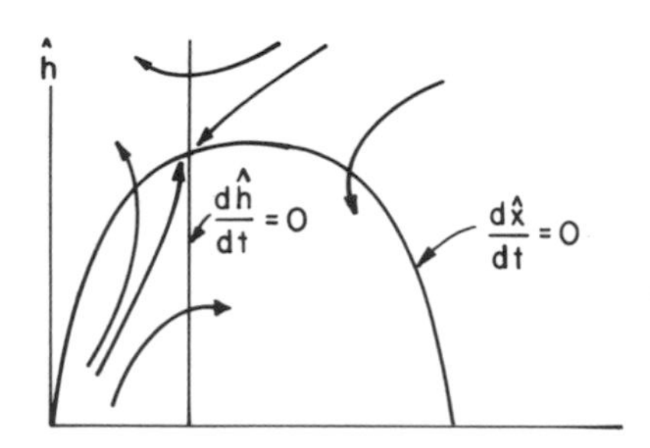

Figure 9.2 Isoclines for Equations 9.40 and 9.41.

we can determine the optimal orbit by means of an isocline analysis. The unique equilibrium $\begin{pmatrix}\tilde{x}\\ \tilde{h}\end{pmatrix}$ corresponding to these equations is defined by $f'(\hat{x}) = \delta$ and $f(\hat{x}) = \hat{h}$. Since $f(x)$ is logistic and $R'/R'' < 0$, the isoclines reveal a saddle point (Figure 9.2). Indeed the Jacobian of 9.40 and 9.41 about $\begin{pmatrix}\tilde{x}\\ \tilde{h}\end{pmatrix}$ is

$$\left.\begin{pmatrix} f' & -1 \\ -\dfrac{R'f''}{R''} & (\delta - f')\left(\dfrac{R''}{R''} - \dfrac{R'}{R'''}\right) \end{pmatrix}\right|_{\begin{pmatrix}\tilde{x}\\ \tilde{h}\end{pmatrix}} = \begin{pmatrix} \delta & -1 \\ -\dfrac{R'f''}{R''} & 0 \end{pmatrix}.$$

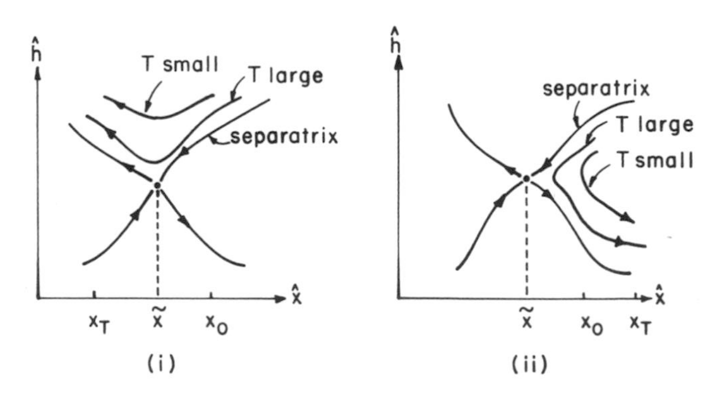

Figure 9.3 Typical orbits linking x_0 to x_T optimally. In (i) x_T is less than $\tilde{x}$ which is less than x_0, while in (ii) x_T is greater than x_0, which is greater than $\tilde{x}$. As T increases, the orbits get closer to the equilibrium and slow down in the vicinity of this rest point. As T goes to infinity, the optimal path lies along a separatrix and the equilibrium is approached asymptotically. In this case, x_T is irrelevant.

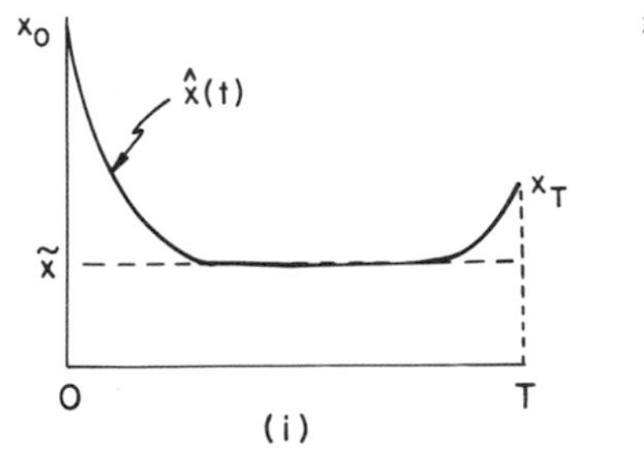

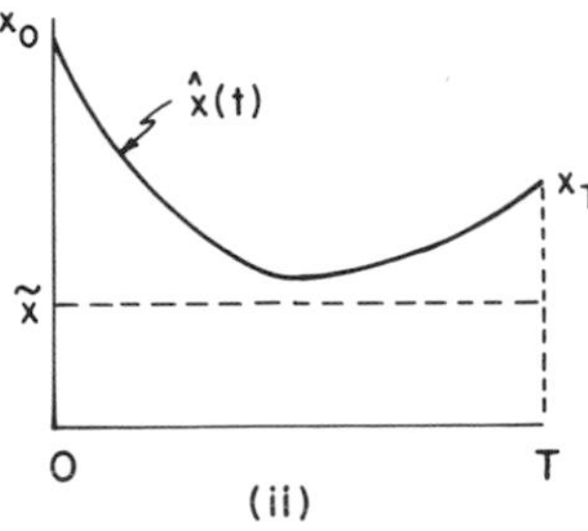

Figure 9.4 Comparison of (i) linear and (ii) catenary "turnpikes" in the case in which x_0 and x_T are both greater than $\tilde{x}$.

The trace of this matrix is positive and the determinant negative, which verifies that the equilibrium is a saddle.

The optimal orbit joins x_0 to x_T. The closer it gets to the equilibrium (where the derivatives of $\hat{x}$ and $\hat{h}$ are zero) the longer it takes. Indeed, along a separatrix it requires an infinite amount of time. Figure 9.3 displays several of the possibilities.

It is instructive to compare optimal paths for revenues that are linear in h and those that are nonlinear. In the first instance, one obtains a linear turnpike, as observed earlier, but in the other case the path approaches the turnpike asymptotically before veering off. We call this a *catenary turnpike*. These are shown in Figure 9.4 for the case in which x_0 and x_T are both greater than $\tilde{x}$.

9.3 Bang–Bang Controls

Whenever control functions are required to satisfy a constraint of the form $\alpha \leq u(t) \leq \beta$, there is the possibility that an optimal $\hat{u}$ switches back and forth between the two values α and β. In this case, $\hat{u}$ is said to be *bang–bang*. In the fish harvesting example of the preceding section, $\hat{u}$ is bang–bang only under exceptional circumstances since $\sigma(t)$ is typically zero in some nontrivial subinterval. This forces $\hat{u}$ to take on an intermediate value as well. But in other problems, it is sometimes possible to deduce that an optimum is indeed bang–bang. This idea is illustrated in the subsequent examples.

Example 9.2 Let us return to the mass–spring system of Chapter One, whose equation of motion is given as

$$\ddot{z} + \omega^2 z + r\dot{z} = 0. \tag{9.42}$$

The state of rest occurs at $z = \dot{z} = 0$, and provided that $r > 0$, an initial disturbance from this position is eventually damped to zero. However, r may be small enough that the return to rest happens quite slowly. In physical applications such sluggish response may be unacceptable. For instance, z could denote the position of some flexible surface on an aircraft that is to be maintained at rest. Occasional wind gusts displace it however, and so the surface behaves approximately like a damped harmonic oscillator.

By imposing a restoring force in the form of a control u, it may be possible to insure that z and $\dot{z}$ both go to zero quickly. In fact, our goal is to choose u so that $\mathbf{x} = \begin{pmatrix} z \\ \dot{z} \end{pmatrix}$ decays to zero in minimum time. Since the restoring force is actuated by a physical device, u cannot be unlimited. Suppose, for example, that the necessary motion is provided by an electric motor in which a change in voltage regulates the magnitude of the movement, with a positive voltage giving rise to an adjustment of the surface in one direction and a negative voltage doing the opposite. The voltage remains within bounds, and so we write

$$|u(t)| \leq \gamma \tag{9.43}$$

for some constant γ. Equation 9.42 is now expressed as

$$\ddot{z} + r\dot{z} + \omega^2 z = u$$

or, as a first-order system,

$$\begin{aligned} \dot{x}_1 &= x_2 \\ \dot{x}_2 &= -\omega^2 x_1 - r x_2 + u \end{aligned}$$

in which $x_1 = z$ and $x_2 = \dot{z}$. For simplicity, we assume that natural damping is so slight that r can be taken to be zero. In this case, the need for some control u is evidently quite essential if the system is not to oscillate indefinitely. Our equations are therefore

$$\begin{aligned} \dot{x}_1 &= x_2 \\ \dot{x}_2 &= -\omega^2 x_1 + u. \end{aligned} \tag{9.44}$$

Initially, $\mathbf{x}(0) = \mathbf{x}_0$, and our problem is to minimize the time T at which $\mathbf{x}$ is first zero. That is, a control u is sought to minimize the integral

$$\int_0^T dt = \int_0^T f_0(\mathbf{x}, u)\, dt$$

with $\mathbf{x}(T) = \mathbf{0}$ and $-\gamma \le u(t) \le \gamma$. The time T is of course unspecified.

Define a function H by

$$H(\mathbf{x}, \lambda, u) = \lambda_0 + \lambda_1 x_2 - \omega^2 \lambda_2 x_1 + u\lambda_2.$$

We know that the λ_i satisfy the equations

$$\dot{\lambda}_1 = -\frac{\partial H}{\partial x_1} = \omega^2 \lambda_2$$

$$\dot{\lambda}_2 = -\frac{\partial H}{\partial x_2} = -\lambda_1.$$

These two relations may be combined into a second-order single equation by differentiating λ_2. This gives $\ddot{\lambda}_2 + \omega^2\lambda_2 = 0$, whose solution is easily obtained in one of several equivalent forms as in Chapter One. The most convenient form of the solution for our purposes is

$$\lambda_2 = B \sin(\omega t - \phi)$$

in which B is a positive "magnitude" and $\phi > 0$ is a suitable "phase angle".

Since H is to be maximized, it is clear that u should be as large as possible if the sign of λ_2 is positive and as small as possible otherwise. Because u is bounded in magnitude by γ, we obtain

$$\hat{u} = \gamma \operatorname{sgn} \lambda_2 = \gamma \operatorname{sgn}(\sin(\omega t - \phi)). \tag{9.45}$$

Thus, $\hat{u}$ is alternately plus and minus γ on intervals of length π/ω, with the first full interval beyond zero beginning at $t = \phi/\omega$. It follows that $\hat{u}$ is a bang–bang control and is implemented in practice by a physical device that switches back and forth between two positions. For example, the voltage regulator mentioned above can be varied between its two extreme values on a regular basis by use of a flip–flop switch.

The orbits of 9.44 corresponding to the optimal bang–bang controller may now be determined. Suppose that $\hat{u} = \gamma$. It is easy to see that x_1

satisfies a second-order equation of the form $\ddot{p} + f(p) = 0$ in which "energy" is conserved. From the phase plane analysis of Section 2.4, we know that

$$\frac{(\dot{x}_1)^2}{2} + \frac{\omega^2 x_1^2}{2} - \gamma x_1 = \text{constant}.$$

Let $v = x_1/\omega$. Then this relation becomes

$$\dot{v}^2 + \left(v - \frac{\gamma}{\omega}\right)^2 = \text{constant},$$

and this defines a family of circles about γ/ω. If $\hat{u} = -\gamma$, one obtains in a similar way circles about $-\gamma/\omega$. Since $\dot{x} > 0$ as x increases, the orbits are clockwise and are displayed in Figure 9.5. It is not hard to see that the time required to complete a half circle is π/ω.

Assuming that the zero position in the $v, \dot{v}$ phase plane is reached at some time T, it is apparent that $\hat{u}$ is either γ or $-\gamma$ in the final switch. Suppose, without loss of generality, that $\hat{u} = \gamma$. The orbit is then circular about γ/ω, and since it reaches the origin in a time not exceeding π/ω in this final segment, it must begin on a semicircle of radius γ/ω in the lower half plane at some point a_1, as shown in Figure 9.6.

Prior to reaching a_1 the orbit is determined by a control $\hat{u} = -\gamma$ for a duration π/ω in which it completes a semicircle about the point $-\gamma/\omega$. This means that it must begin this leg of its journey at a point a_2 on a line through $-\gamma/\omega$ and joining a_1 (Figure 9.7).

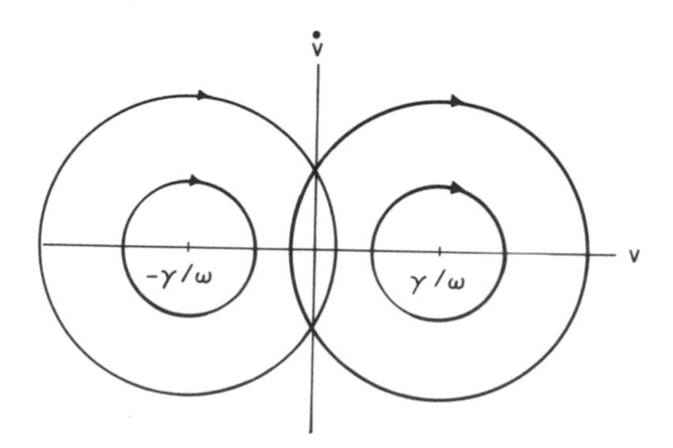

Figure 9.5 Two families of circular orbits in the $v, \dot{v}$ phase plane for the equations $\ddot{x}_1 + \omega^2 x_1 \pm \gamma = 0$ when $v = x_1/\omega$.

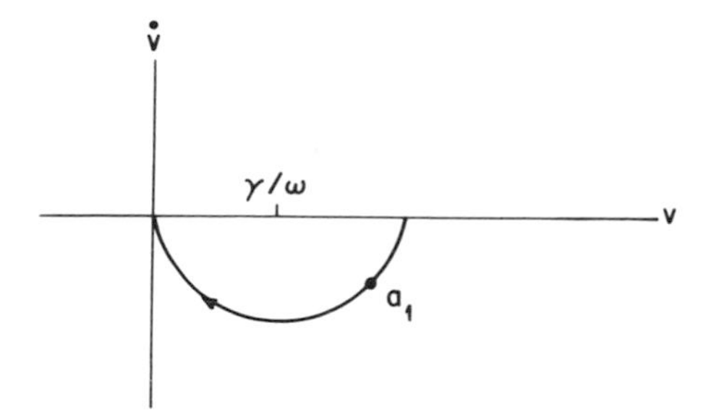

Figure 9.6 The last stage of an optimal path in the $v, \dot{v}$ phase plane in which the origin is reached via the control $\hat{u} = \gamma$, starting at a_1 on the semicircle.

Proceeding retrogressively through a finite number of semicircular orbits, we eventually reach the initial interval of duration ϕ/ω in which $\hat{u}$ is either γ or $-\gamma$. A virtually identical argument applies to the case in which $\hat{u} = -\gamma$ in the final switch, and this results in an arrangement of orbits which are symmetric to the ones just found. By combining both pictures into one, we obtain Figure 9.8.

In Figure 9.8, we see a periodic curve C that resembles a ruffle that is attached to the horizontal axis. From the discussion above, we note that $\hat{u}$ is γ for all points in the $v, \dot{v}$ plane that lie below C, while above C it takes on the value $-\gamma$. We conclude that $\hat{u}$ depends on v and $\dot{v}$ (and

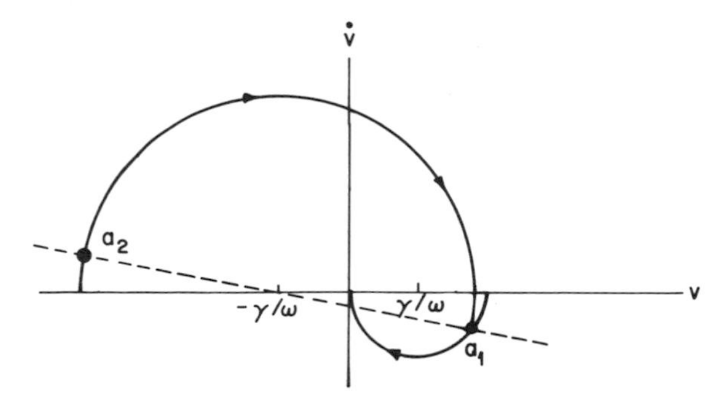

Figure 9.7 The last two stages of an optimal path in the $v, \dot{v}$ phase plane. The control $\hat{u}$ is $-\gamma$ along the semicircle beginning at a_2 until it switches to γ when it reaches a_1. The semicircle containing a_2 has a diameter which joins a_2 to a_1 through a center at $-\gamma/\omega$.

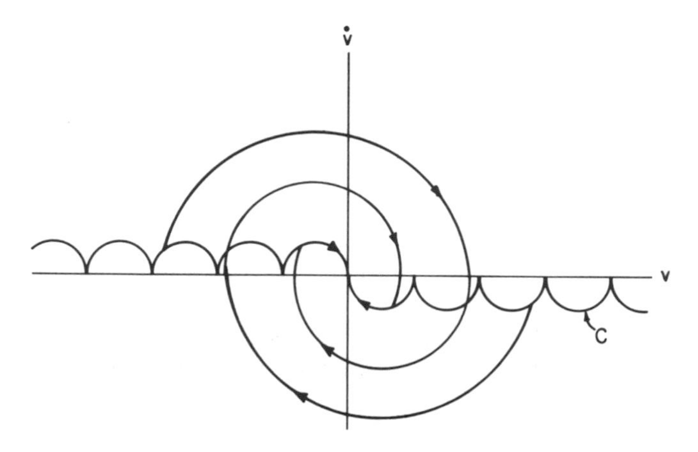

Figure 9.8 Two families of optimal orbits, depending on whether the final value of $\hat{u}$ is γ or $-\gamma$. All motion above the curve C is defined by $\hat{u} = -\gamma$ and by $\hat{u} = \gamma$ below C.

therefore on x_1 and $\dot{x}_1$) as a feedback control:

$$\hat{u}(t) = \begin{cases} \gamma & \text{in the } v, \dot{v} \text{ plane below } C, \\ -\gamma & \text{in the } v, \dot{v} \text{ plane above } C. \end{cases}$$

Since v and $\dot{v}$ are scalar multiples of x_1 and $\dot{x}_1$, the optimal control law can bè rewritten as some two-valued function $\hat{u} = g(x_1, \dot{x}_1)$ in the $x_1, \dot{x}_1$ plane. We can interpret it as a dry or Coulomb frictional force that brings the system to rest quickly in a series of jerky motions. If u did not depend on x_1 but only on $\dot{x}_1 = x_2$ as in the standard model of a mass–spring with dry friction, then from our earlier discussion in Chapter One, we would have

$$u = g(\dot{x}_1) = \begin{cases} \gamma & \text{if } \dot{x}_1 < 0, \\ -\gamma & \text{if } \dot{x}_1 > 0. \end{cases}$$

In this case, the control brings the system to the horizontal axis, where the velocity $\dot{x}_1$ is zero, but not necessarily to the position $x_1 = 0$ (recall Exercise 2.5.8 and consult Figure 9.9). Thus, it is not optimal, and we see that an optimal control must be a function of both x_1 and $\dot{x}_1$.

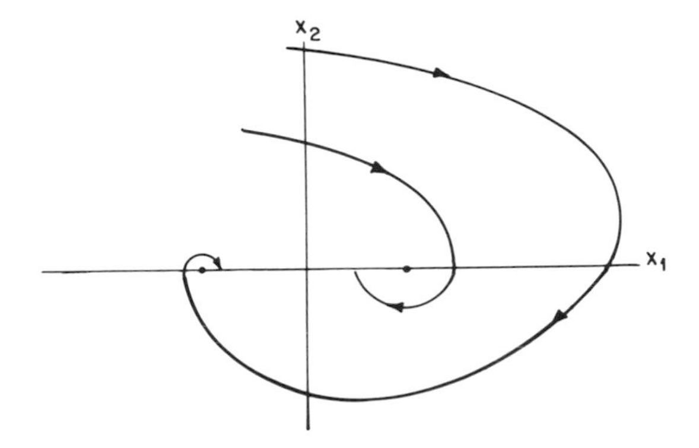

Figure 9.9 Orbits corresponding to a nonoptimal control $\hat{u}$ that depends only on $\dot{x}_1$. Motion is brought to a halt but not necessarily at the desired position $x_1 = 0$.

Example 9.3 (Rocket Motion) Consider a rocket moving in a plane perpendicular to the earth. It accelerates by burning fuel that is expelled out of the rear of the rocket as an exhaust gas. The force exerted by the exhaust on the rocket (equal and opposite to the force exerted by the rocket on the gas to expel it) propels the rocket forward.

Initially, at some time t, the rocket has velocity v and mass m. During a subinterval of duration Δt it burns fuel whose total mass is Δm. The ejected gas has a velocity $\hat{v}$ relative to the rocket. At time $t + \Delta t$, the rocket has mass $m - \Delta m$ and velocity $v + \Delta v$, while the ejected fuel has a velocity $v + \hat{v}$ and mass Δm. In the absence of any external forces, the rocket plus expelled fuel satisfies the law of conservation of momentum. Initially, the momentum is mv, but at time $t + \Delta t$ it is

$$(m - \Delta m)(v + \Delta v) + \Delta m(v + \hat{v}).$$

The conservation law implies that

$$mv - (m - \Delta m)(v + \Delta v) - \Delta m(v + \hat{v}) = 0. \tag{9.46}$$

Suppose that fuel is burned at a rate $\beta(t)$ that is the limit of $\Delta m/\Delta t$ as Δt tends to zero. Divide 9.46 by Δt, cancel out like terms, and then let Δt go to zero. This results in

$$m\dot{v} + \beta\hat{v} = 0. \tag{9.47}$$

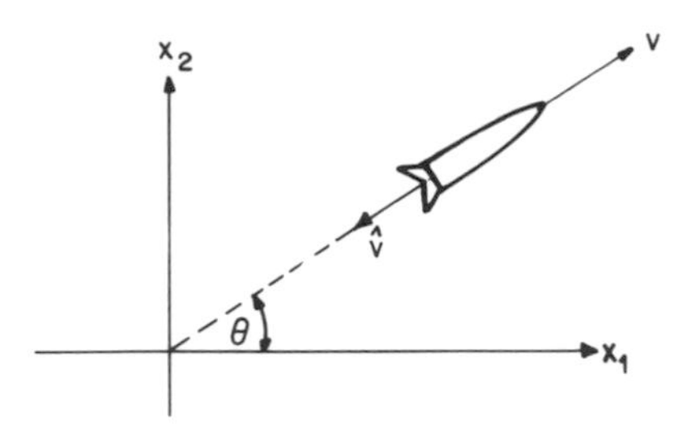

Figure 9.10 Schematic of rocket motion in which the exhaust gas is expelled from the rear of the rocket at a velocity $\hat{v}$ relative to that of the rocket. $\hat{v}$ has components $\mu \cos \theta$ and $\mu \sin \theta$ along the x_1 and x_2 axis. θ is called the "thrust angle".

If the rocket moves near the earth, there is a constant gravitational force g per unit mass to contend with and 9.47 must be modified. In order to incorporate this, let $\mathbf{x} = \begin{pmatrix} x_1 \\ x_2 \end{pmatrix}$ denote the horizontal and vertical coordinates of the rocket relative to the plane that is perpendicular to the earth and $\mathbf{v} = \begin{pmatrix} v_1 \\ v_2 \end{pmatrix}$ the corresponding velocity components. Define θ to be the angle of motion relative to the x_1 axis (Figure 9.10). Since the change in momentum can only be due to the gravitational force, which acts along the x_2 axis, the right side of 9.47 must equal $\begin{pmatrix} 0 \\ -mg \end{pmatrix}$.

Assume that the exhaust velocity has a constant magnitude μ. Only its direction varies, and so the components of $\hat{v}$ are $-\mu \cos \theta$ and $-\mu \sin \theta$. Combining all this information with 9.47 yields the differential equations

$$\begin{aligned}
\dot{x}_1 &= v_1 \\
\dot{x}_2 &= v_2 \\
\dot{v}_1 &= \frac{\beta\mu \cos \theta}{m} \\
\dot{v}_2 &= \frac{\beta\mu \sin \theta}{m} - g \\
\dot{m} &= -\beta.
\end{aligned} \tag{9.48}$$

Note the negative sign in the last equation. The reason this comes about is that $m(t + \Delta t) - m(t) = m - \Delta m - m = -\Delta m$.

The *thrust* of the rocket is defined by $\mu\beta$. The rocket's acceleration is due, in effect, to the thrust of the engine as a result of burning fuel and expelling the exhaust. The thrust is a force whose magnitude is proportional to β and a direction determined by the *thrust angle* θ.

Physical considerations limit the burning rate as well as the thrust angle. We express this as inequality constraints

$$\begin{aligned} 0 \le \beta(t) \le \bar{\beta} \\ 0 \le \theta(t) \le \bar{\theta}. \end{aligned} \tag{9.49}$$

Both β and θ are to be thought of as control variables.

Consider a special case in which the rocket is to move a fixed nonzero payload of mass $\overline{m}$ vertically upward. We wish to choose a burning rate β that allows the rocket to reach a maximum altitude. Since motion is straight up, the thrust angle takes on the value $\theta(t) = \pi/2$. Also, the horizontal axis plays no role, and so Equations 9.48 and constraints 9.49 are simplified. All the fuel in the rocket is to be used up during burning, and the final velocity is left unspecified. We leave it to the reader to complete the mathematical formulation of this problem and to show that the optimal burning rate β is actually bang–bang with only one switch in value (Exercise 9.4.4).

9.4 Exercises

9.4.1 Show that the system of equations

$$-\dot{K} = KA + A^{\mathrm{T}}K + I - K\mathbf{b}\mathbf{b}^{\mathrm{T}}K = 0.$$

(Equation 9.18 in the text) has a unique positive definite and symmetric solution K that is constant for all t, provided r is small enough.

9.4.2 Establish that the scalar λ_0 in 9.37 is nonzero if one is to avoid a contradiction of Theorem 9.1.

9.4.3 Show that the quadratic equation 9.33 has two real roots, one of which is positive.

9.4.4 Complete the formulation of the vertical rocket flight problem that was begun in Example 9.3. Then establish the optimal control to be of bang–bang type in which burning takes place initially at a maximum rate ("full thrust") until all the fuel is used up, and then the engine shuts off allowing the rocket to coast the rest of the way until it attains its maximum altitude.

9.4.5 Consider an open access fishery whose dynamics are captured by Equations 9.21 and 9.23. Suppose that a regulatory agency imposes a tax τ per unit catch of fish. This obliges us to revise 9.23. Assume that $\tau_1 \le \tau(t) < \tau_2$ with $\tau_2 > 0$. If τ_1 is less than zero, it means that τ can represent a subsidy as well as a tax, depending on its sign. Letting τ be the only control formulate a mathematical model for maximizing the discounted net income over some given time interval.

9.4.6 Using Theorem 9.1 obtain a description of the optimal tax policy for the problem in Exercise 9.4.5.

9.4.7 A special case of the rocket model of Example 9.3 is that of a rocket moving between two fixed locations in different orbits about the earth. Assume that the thrust $\mu\beta$ is constant and that there is no constraint on the thrust angle θ. Formulate the mathematical problem of minimum time rendezvous between the two given orbits. Show that the optimum θ is of the form

$$\hat{\theta}(t) = \frac{\alpha t - \beta}{\gamma t - \delta}$$

for suitable constants. What happens to $\hat{\theta}$ if the final velocity is not specified?

9.4.8 Consider a simple form of the predator–prey model given by

$$\begin{aligned} \dot{x}_1 &= x_1(1 - x_2) \\ \dot{x}_2 &= -x_2(1 - x_1). \end{aligned}$$

To be specific, we think of x_1 as the density of a crop pest and x_2 the density of its natural predator. An insecticide is applied that controls the level of x_1, but which does not harm the predator. The insecticide is sprayed at a rate $u(t)$ per unit prey density. The pest is destroyed at a rate proportional to the spray rate.

Without the external control, the values of x_1 and x_2 fluctuate about an equilibrium, as we know from the discussion of this model in Chapter Four. We would like u to be chosen so that the populations of both pest and predator reach equilibrium at some time T. The maximum spray rate is of course limited. The cost of applying the spray over the given time interval, which is proportional to the total amount of insecticide used, should be as low as possible. The total population of the pest during this interval should also be minimized. Formulate this as an optimal control problem and solve. Show that the optimal spray program $\hat{u}$ is bang–bang.

Appendix

Ordinary Differential Equations: A Review

What follows is a brief review of the salient facts concerning ordinary differential equations as needed in the text.

We are given $k \leq 3$ equations of the form

$$\dot{x}_i = f_i(x, t), \qquad 1 \leq i \leq k, \tag{A.1}$$

where f_i are usually nonlinear functions in x and t. In vector notation, A.1 may be written as

$$\dot{\mathbf{x}} = \mathbf{f}(\mathbf{x}, t) \tag{A.2}$$

in which $\mathbf{x}$ and $\mathbf{f}$ are vectors in $\mathbf{R}^k$ having components x_i and f_i respectively.

It is customary to specify the initial conditions $\mathbf{x}(0) = \mathbf{x}_0$. The function $\mathbf{f}$ is defined for all $\mathbf{x}$ in some open set U in $\mathbf{R}^k$ and is assumed to be smooth (that is, at least continuously differentiable) with respect to $\mathbf{x}$ and t. This guarantees that for every $\mathbf{x}_0$ in U there is a unique solution of A.2 passing through $\mathbf{x}_0$ at time $t = 0$. More specifically, there is a uniquely defined vector function $\boldsymbol{\phi}(t)$ and some interval $|t| < \delta$ for which $\boldsymbol{\phi}(0) =$

x_0 and $\dot{\boldsymbol{\phi}}(t) = \mathbf{f}(\boldsymbol{\phi}(t), t)$. In the applications given in the text, δ is assumed to be infinite.

The set U is generally unspecified, but it is understood to include all states of physical interest.

First-Order Equations (The Case $k = 1$)

Consider the scalar equation $\dot{x} = f(x, t)$ in which f can be written as a quotient $f_1(t)/f_2(x)$. That is,

$$f_2(x)\dot{x} = f_1(t). \tag{A.3}$$

This is called a *separable equation*. By the usual change of variable formula in calculus,

$$\int_0^t f_L(x(t)) \frac{dx}{dt}\, dt = \int_{x_0}^{x(t)} f_2(x)\, dx.$$

By a formal integration of A.3 one obtains

$$\int_{x_0}^{x(t)} f_2(x)\, dx = \int_0^t f_1(s)\, ds. \tag{A.4}$$

If f_1 and f_2 are simple enough, A.4 can be explicitly integrated to give $x(t)$. One of the simplest examples is the linear equation

$$\dot{x} = a(t)x. \tag{A.5}$$

It is easy to see from A.4 that the solution to A.5 is

$$x(t) = x(0)\, e^{A(t)}, \tag{A.6}$$

where

$$A(t) = \int_0^t a(s)\, ds.$$

In the special case in which a is a constant, the solution A.6 reduces to

$$x(t) = x(0)\, e^{at}.$$

The nonhomogeneous version of A.5 is

$$\dot{x}(t) = a(t)x + b(t). \tag{A.7}$$

If x is a solution of A.7, let

$$u(t) = e^{-A(t)}x(t)$$

with $A(t)$ as defined above. Then we see that

$$\dot{x} - a(t)x = e^{A(t)}\dot{u}.$$

By virtue of A.7 this gives

$$\dot{u} = e^{-A(t)}b(t),$$

and so

$$u(t) = \int_0^t e^{-A(s)}b(s)\,ds + u(0).$$

It follows that the solution to A.7 is

$$x(t) = e^{A(t)}\left[x(0) + \int_0^t e^{-A(s)}b(s)\,ds\right] \tag{A.8}$$

Let x_p be any particular solution to A.7. Often it can be found by inspection. If x is the general solution, then $x_c = x - x_p$ satisfies the homogeneous equation in which $b = 0$. We saw above that the homogeneous equation has A.6 as its solution. Relation A.8 indeed expresses x as a sum of x_c and some particular solution in a formal way.

The Case $k = 2$

Here we assume that **f** is not explicitly dependent on time. In this case, A.1 is called an *autonomous* system. That is, we have the system

$$\dot{x}_1 = f_1(x_1, x_2)$$
$$\dot{x}_2 = f_2(x_1, x_2)$$

or, in vector notation, $\dot{\mathbf{x}} = \mathbf{f}(\mathbf{x})$ with $\mathbf{x}$ and $\mathbf{f}$ in $\mathbf{R}^2$. An important special case is the linear system

$$\begin{aligned} \dot{x}_1 &= a_{11}x_1 + a_{22}x_2 \\ \dot{x}_2 &= a_{21}x_1 + a_{22}x_2 \end{aligned} \tag{A.9}$$

or, in vector notation,

$$\dot{\mathbf{x}} = A\mathbf{x}$$

in which A is the constant coefficient matrix

$$A = \begin{pmatrix} a_{11} & a_{12} \\ a_{21} & a_{22} \end{pmatrix}.$$

By analogy with the solution to a scalar first-order equation $\dot{x} = ax$ we attempt a solution of the form

$$\mathbf{x}(t) = \mathbf{c}\, e^{\lambda t},$$

where λ is a scalar and $\mathbf{c}$ is a nonzero vector $\begin{pmatrix} c_1 \\ c_2 \end{pmatrix}$. That is, one tries to form a vector solution $\mathbf{x}(t)$ whose components satisfy

$$\begin{pmatrix} x_1(t) \\ x_2(t) \end{pmatrix} = e^{\lambda t} \begin{pmatrix} c_1 \\ c_2 \end{pmatrix}. \tag{A.10}$$

Substitute A.10 into A.9 to obtain

$$A\mathbf{c} = \lambda \mathbf{c}. \tag{A.11}$$

A nontrivial solution to this relation for a given λ is called an *eigenvector* and λ is the corresponding *eigenvalue*. Since the vector $\mathbf{c}$ is required to be a nonzero solution to a homogeneous algebraic system, it is necessary that

$$\mathrm{Det}(A - \lambda I) = 0.$$

This determinant equation gives rise to the quadratic *characteristic equation*

$$\lambda^2 - \lambda\, \mathrm{Trace}\, A + \mathrm{Det}\, A = 0,$$

which has either two real roots or a pair of complex conjugate roots. Assume these roots λ_1 and λ_2 are distinct. This is the usual or generic case, as explained in Chapter Three. It follows as a simple exercise that the corresponding eigenvectors $\mathbf{c}_1 = \begin{pmatrix} c_{11} \\ c_{12} \end{pmatrix}$ and $\mathbf{c}_2 = \begin{pmatrix} c_{21} \\ c_{22} \end{pmatrix}$ are linearly independent vectors in $\mathbf{R}^2$. They are determined by solving A.11 for each λ_i.

To illustrate this consider the case in which

$$A = \begin{pmatrix} 0 & 1 \\ 1 & 0 \end{pmatrix}.$$

The eigenvalues are ± 1 and so the corresponding eigenvectors are easily computed to be $\begin{pmatrix} 1 \\ 1 \end{pmatrix}$ and $\begin{pmatrix} 1 \\ -1 \end{pmatrix}$.

Since $\mathbf{c}_i e^{\lambda_i t}$ are both solutions, for $i = 1, 2$, so is any linear combination of them (because of the linearity of A.9). In fact, the most general solution is of the form since every vector in $\mathbf{R}^2$ can be written as a linear combination of two linearly independent quantities. Therefore,

$$\mathbf{x}(t) = \alpha_1 e^{\lambda_1 t} \mathbf{c}_1 + \alpha_2 e^{\lambda_2 t} \mathbf{c}_2, \tag{A.12}$$

where α_1 and α_2 are suitable scalars determined by a knowledge of $\mathbf{x}(0)$. Thus, if $t = 0$, then A.12 gives

$$\begin{pmatrix} c_{11} & c_{12} \\ c_{21} & c_{22} \end{pmatrix} \begin{pmatrix} \alpha_1 \\ \alpha_2 \end{pmatrix} = \begin{pmatrix} x_1(0) \\ x_2(0) \end{pmatrix},$$

which has a unique solution in α_1, α_2 since $\mathbf{c}_1, \mathbf{c}_2$ are linearly independent.

Example A.1 An important special case is the linear second-order equation

$$\ddot{y} + a\dot{y} + by = 0 \tag{A.13}$$

in which a, b are constants. Setting $x_1 = y$ and $x_2 = \dot{y}$ gives

$$\begin{aligned} \dot{x}_1 &= \phantom{-bx_1 - {}} x_2 \\ \dot{x}_2 &= -bx_1 - ax_2, \end{aligned}$$

which is an instance of A.9. Since

$$A = \begin{pmatrix} 0 & 1 \\ -b & -a \end{pmatrix}$$

in this case, the eigenvalues are computed from the quadratic equation

$$\lambda^2 + a\lambda + b = 0.$$

The solution to A.13 is given by $y = x_1$. For example, if $a = 0$ and $b = -1$ in A.13, then as computed above, the eigenvalues are ± 1 with eigenvectors $\begin{pmatrix} 1 \\ 1 \end{pmatrix}$ and $\begin{pmatrix} 1 \\ -1 \end{pmatrix}$. From A.12, one obtains

$$x(t) = \alpha_1 e^{t} \begin{pmatrix} 1 \\ 1 \end{pmatrix} + \alpha_2 e^{-t} \begin{pmatrix} 1 \\ -1 \end{pmatrix},$$

so that

$$y(t) = x_1(t) = \alpha_1 e^t + \alpha_2 e^{-t}.$$

An interesting situation occurs when the roots λ_i in A.12 are complex:

$$\lambda_1 = \sigma + iq$$
$$\lambda_2 = \sigma - iq$$

with $q \neq 0$ (since we assume distinct roots of the characteristic equation). In physical applications, the solution $x(t)$ is generally required to be real. Since the expression in A.12 could, as its stands, be complex it is expedient to ask that $x(t) = \overline{x(t)}$, where the bar indicates complex conjugation. That is, $\begin{pmatrix} x_1 \\ x_2 \end{pmatrix} = \begin{pmatrix} \bar{x}_1 \\ \bar{x}_2 \end{pmatrix}$. Because

$$e^{\lambda t} = e^{\sigma t} e^{\pm iqt},$$

this translates into (with $\boldsymbol{\beta}_i = \alpha_i \mathbf{c}_i$)

$$e^{\sigma t}\left(\boldsymbol{\beta}_1 e^{iqt} + \boldsymbol{\beta}_2 e^{-iqt}\right) = e^{\sigma t}\left(\bar{\boldsymbol{\beta}}_1 e^{-iqt} + \bar{\boldsymbol{\beta}}_2 e^{iqt}\right),$$

which implies that

$$e^{2iqt}\left(\boldsymbol{\beta}_1 - \bar{\boldsymbol{\beta}}_2\right) = \bar{\boldsymbol{\beta}}_1 - \boldsymbol{\beta}_2.$$

This relation forces $\boldsymbol{\beta}_2$ to equal $\bar{\boldsymbol{\beta}}_1$. Let $\boldsymbol{\beta}_2 = \boldsymbol{\zeta} + i\boldsymbol{\eta}$. Then $\boldsymbol{\beta}_1 = \boldsymbol{\zeta} - i\boldsymbol{\eta}$, and from A.12,

$$\mathbf{x}(t) = e^{\sigma t}[(\boldsymbol{\zeta} - i\boldsymbol{\eta})(\cos qt + i\sin qt) + (\boldsymbol{\zeta} + i\boldsymbol{\eta})(\cos qt - i\sin qt)]$$

in which we have used the identity

$$e^{\pm iqt} = \cos qt \pm i\sin qt.$$

It follows that

$$\mathbf{x}(t) = e^{\sigma t}(\boldsymbol{\zeta}_1 \cos qt + \boldsymbol{\zeta}_2 \sin qt), \tag{A.14}$$

where the vectors $\boldsymbol{\zeta}_1$ and $\boldsymbol{\zeta}_2$ are given by

$$\boldsymbol{\zeta}_1 = \boldsymbol{\beta}_1 + \boldsymbol{\beta}_2 \qquad \text{and} \qquad \boldsymbol{\zeta}_2 = \frac{1}{i}(\boldsymbol{\beta}_1 - \boldsymbol{\beta}_2).$$

Example A.2 Consider the equation in Example A.1 in which $a = 0$. Then the eigenvalues are $\pm i\sqrt{b}$. The corresponding eigenvectors are $\begin{pmatrix} 1 \\ i\sqrt{b} \end{pmatrix}$ and $\begin{pmatrix} 1 \\ -i\sqrt{b} \end{pmatrix}$, so that $\boldsymbol{\beta}_1 = \alpha_1\begin{pmatrix} 1 \\ i\sqrt{b} \end{pmatrix}$ and $\boldsymbol{\beta}_2 = \alpha_2\begin{pmatrix} 1 \\ -i\sqrt{b} \end{pmatrix}$. Since $\sigma = 0$, it follows from A.14 that $y = x_1$ is given by

$$y(t) = (\alpha_1 + \alpha_2)\cos\sqrt{b}\,t - i(\alpha_1 - \alpha_2)\sin\sqrt{b}\,t.$$

To determine α_1 and α_2, we need to know $\mathbf{x}(0)$, namely $y(0)$ and $\dot{y}(0)$. Suppose, for instance, that $y(0) = 1$ and $\dot{y}(0) = 0$. Then one immediately finds that $\alpha_1 = \alpha_2 = \frac{1}{2}$. Hence,

$$y(t) = \cos\sqrt{b}\,t.$$

Now consider the nonhomogeneous version of A.13

$$\ddot{y} + a\dot{y} + by = \gamma(t), \tag{A.15}$$

where γ is some given function. Let y_p be any *particular solution* to A.15. In all instances in this book, y_p is found by judicious guessing. If y is the most general solution to A.15, then $y_c \equiv y - y_p$ satisfies the homogeneous version of this equation (i.e., in which $\gamma = 0$). We saw above in Example A.1 how to find y_c in all cases. It follows that

$$y = y_c + y_p$$

is the *general solution*.

An important generalization of A.13 is the second-order equation

$$\ddot{y} + g(y) = 0. \tag{A.16}$$

Let $x_1 = y$ and $x_2 = \dot{y}$ as before. Then A.16 becomes a special case of $\dot{\mathbf{x}} = \mathbf{f}(\mathbf{x})$, namely

$$\dot{x}_1 = x_2$$
$$\dot{x}_2 = -g(x_1)$$

in which $\mathbf{f}(\mathbf{x}) = \begin{pmatrix} x_2 \\ -g(x_1) \end{pmatrix}$.

The Case $k = 3$

Here we have $\dot{\mathbf{x}} = \mathbf{f}(\mathbf{x})$ with $\mathbf{x}$ in $\mathbf{R}^3$. If $\mathbf{f}$ is linear in the variables $\mathbf{x}_1$, $\mathbf{x}_2$, and $\mathbf{x}_3$, then $\dot{\mathbf{x}} = A\mathbf{x}$ and one may show as before that the general solution is given as a linear combination

$$\mathbf{x}(t) = \sum_{i=1}^{3} \alpha_i \mathbf{c}_i e^{\lambda_i t}, \tag{A.17}$$

where λ_i are the eigenvalues of A (assumed distinct) and $\mathbf{c}_i$ the corresponding independent eigenvectors. The λ_i are either all real or one is real and the others are a complex conjugate pair.

Example A.3 Suppose $\dot{\mathbf{x}} = A\mathbf{x}$ with

$$A = \begin{pmatrix} 2 & 1 & 1 \\ 2 & 3 & 4 \\ -1 & -1 & -2 \end{pmatrix}.$$

Its characteristic equation is

$$\mathrm{Det}(A - \lambda I) = (\lambda - 1)(\lambda + 1)(\lambda - 3) = 0,$$

so the eigenvalues are 1, -1, and 3. The corresponding eigenvectors are computed to be

$$\begin{pmatrix} 1 \\ -1 \\ 0 \end{pmatrix}, \quad \begin{pmatrix} 0 \\ 1 \\ -1 \end{pmatrix}, \quad \begin{pmatrix} 2 \\ 3 \\ -1 \end{pmatrix}.$$

Hence,

$$x(t) = \alpha_1 e^{t} \begin{pmatrix} 1 \\ -1 \\ 0 \end{pmatrix} + \alpha_2 e^{-t} \begin{pmatrix} 0 \\ 1 \\ -1 \end{pmatrix} + \alpha_3 e^{3t} \begin{pmatrix} 2 \\ 3 \\ -1 \end{pmatrix}.$$

If $\mathbf{x}(0) = \begin{pmatrix} 1 \\ 0 \\ 0 \end{pmatrix}$ then $\alpha_1 = \frac{1}{2}$, $\alpha_2 = -\frac{1}{4}$, $\alpha_3 = \frac{1}{4}$.

References and a Guide to Further Readings

The literature on the mathematics of dynamical system models is huge, and we make no pretence at giving a comprehensive survey. The few references cited below were included for one of several reasons. First, they were of direct relevance to material covered in the text. Second, they offer readable discussions at a level that is accessible to the readers of this book. The more difficult or esoteric references were omitted as well as the more elementary sources.

Here and there the authors and dates of some of the original modeling efforts are also mentioned, solely for historical reasons. It is sobering to know that much of what is of active interest today has its roots in research done long ago.

General references are listed below. Following this we comment on specific sources for each of the Chapters.

Ordinary Differential Equations

Among the many references to this topic the best modern introduction known to us is

[1] M. Hirsch and S. Smale, *Differential Equations, Dynamical Systems, and Linear Algebra*, Academic Press, 1974.

As a sequel to [1] on the subject of structural stability and the local behavior of orbits near a hyperbolic fixed point is

[2] M. Peixoto, "Generic Properties of Ordinary Differential Equations," in *Studies of ODE* (edited by J. Hale), MAA, 1977.

A classic and easy to read reference on the subject of stability and Liapunov functions is

[3] J. LaSalle and S. Lefshetz, *Stability by Liapunov's Direct Method*, Academic Press, 1961.

In connection with [3] the following paper is also useful:

[4] J. LaSalle, "Some Extensions of Liapunov's Second Method," *IEEE Trans. on Circuit Theory*, 1960, 520–527.

Introductions to Differential Equation Modeling

Some useful chapters on population modeling and combat models appear in the book

[5] M. Braun, C. Coleman, and D. Drew (editors), *Differential Equation Models*, Springer–Verlag, 1983.

There is a recent book whose later chapters offer an excellent introduction to some of the topics covered by us:

[6] A. Borrelli and C. Coleman, *Differential Equations, A Modeling Approach*, Prentice–Hall, 1987.

A readable introduction to mechanical models and traffic flow is contained in

[7] R. Haberman, *Mathematical Models*, Prentice–Hall, 1977.

A number of biological models, especially those involving partial differential equations, are surveyed in

[8] D. Jones and D. Sleeman, *Differential Equations and Mathematical Biology*, Allen and Unwin, 1983.

An undergraduate introduction to the modeling of dynamical systems that is of special interest to engineering students because of its use of Laplace transform methods (transfer functions) is

[9] D. Luenberger, *Introduction to Dynamic Systems*, Wiley, 1979.

Reference [9] is useful in connection with our Chapters 3 and 9, while some of the topics in Chapters 6 and 9 are treated in

[10] C. Clark, *Mathematical Bioeconomics*, Wiley, 1976.

There is a translation of an older Russian work that gives numerous examples of dynamical systems in the plane with many details and illustrations:

[11] A. Andronov, A. Witt, and S. Chaiken, *Theory of Oscillators*, Addison–Wesley, 1966.

More Advanced Modeling Books

A good survey of a number of topics covered by us is in

[12] J. Murray, *Lectures on Nonlinear Differential Equation Models in Biology*, Oxford, 1977.

An intermediate level, somewhat older, but still valuable reference on nonlinear oscillations is:

[13] J. Stoker, *Nonlinear Vibrations*, Wiley, 1950.

A more contemporary but also more advanced treatment of dynamical systems is:

[14] J. Guckenheimer and P. Holmes, *Nonlinear Oscillations, Dynamical Systems, and Bifurcations of Vector Fields*, Springer–Verlag, 1983.

Especially recommended is the first chapter of [14].

Hard to Classify

There is a non-mathematical, pictorial, quite stimulating and by no means trivial set of paperback volumes on dynamic models:

[15] R. Abraham and C. Shaw, *Dynamics: The Geometry of Behavior*, Aerial Press, 1984 to the present.

I have only seen the first three volumes of [15] but more are scheduled for release in 1987 and later.

The use of computer graphics to study dynamical systems is an invaluable modeling tool that is essential in the study of more complicated orbit behavior. A very useful introduction that is designed to be used with an IBM PC (and which contains an easy to use diskette) is

[16] H. Kocak, *Differential and Difference Equations through Computer Experiments*, Springer–Verlag, 1986.

Notes on the Individual Chapters

Chapter One

Material on the harmonic oscillator is contained in virtually every undergraduate general physics text. The forced harmonic oscillator is also discussed in Haberman [7]. This is only the simplest case of course. Forced oscillations of nonlinear equations, such as those of Van der Pol and Duffing, lead us to the very frontiers of nonlinear dynamics. This topic is treated in detail in Stoker [13] and with more recent but also more tersely explained results in Guckenheimer and Holmes [14]. A visual and intuitively appealing introduction is contained in Abraham and Shaw [15].

The Coulomb friction model reappears in different ways in Chapters 6 and 9. See also Exercise 2.5.8.

We have avoided transform techniques in this text. However, their use does simplify the treatment of some models. In addition to Luenberger [9] the following book also employs these methods: A. McClamroch, *State Models of Dynamic Systems*, Springer–Verlag, 1980. It contains numerous examples of models in which the conservation of mass argument is invoked.

For a discussion of stable equilibria, with examples, see LaSalle and Lefschetz [3].

Chapter Two

For a study of the pendulum equation and phase plane methods, we refer again to Stoker [13] as well as to Haberman [7] and Andronov *et al.* [11]. The last named book has a nice picture of the cylindrical phase space representation of the pendulum. The phase plane analysis of conservative systems reappears in several places, especially in Chapter 5.

The stability of rest states for nonlinear systems has its roots in the seminal work of Poincaré and Liapunov in the 19th century. The linearization theorem is proven in Hirsch and Smale [1] while the more difficult Hartman–Grobman Theorem for hyperbolic fixed points is discussed in Peixoto [2]. This theorem and the technically more difficult stable manifold theorem are also proven in a book that more mathematically prepared readers may find useful: J. Palis and W. deMelo, *Geometric Theory of Dynamical Systems*, Springer–Verlag, 1982.

Chapter Three

Liapunov's theorem, with several examples, is covered in LaSalle and Lefschetz [3], Hirsch and Smale [1], and LaSalle [4]. This result reappears in several places again, notably in Chapters 4, 6, 7, and 9.

For a general discussion of structural stability, an idea that began with Andronov and Pontryagin in 1937 and reached a plateau with Peixoto in 1962, see Hirsch and Smale [1], Peixoto [2], as well as a delightful article by Smale: "What Is Global Analysis?", *Am. Math. Mo.* **76**, 1969, 4–9. Proofs that it is generic for eigenvalues of matrices to be distinct and to possess nonzero real parts is also found in Hirsch and Smale [1].

Feedback control was given its first mathematical treatment in the 19th century in the work of Maxwell in 1868 and Vishnegradski in 1876. A major impetus came from the practical question of stabilizing the Watt flyball governor for engines, a model of which is treated in detail in Chapter 6. A collection of some of the basic papers in the field is given by R. Bellman and R. Kalaba: *Selected Papers in Control Theory*, Dover, 1964.

Feedback as a metaphor for dynamical systems is eloquently described in the classic work, *Cybernetics*, by N. Wiener, John Wiley, 1948.

A proof of the eigenvalue placement theorem and the related ideas of controllability and observability are found in Luenberger [9].

Chapter Four

The pioneering work on quadratic population models is that of Volterra in 1926. An extensive overview of this work, with applications, is contained in a book by U. D'Ancona: *The Struggle for Existence*, Brill (Leiden), 1954. A classic source for the ecological background of population models is L. Slobodkin: *Growth and Regulation of Animal Populations*, 1961, Dover reprint, in which one finds reference to the early work of Lotka (1931) and Gause (1934), the latter for his competitive exclusion principle.

Combat models were first formulated in 1916 by Lanchester, while the classic epidemic model is by Kermac and McKendrick in 1927 (Exercise 4.5.12). A nice exposition of this material from an elementary point of view is in Chapters 8, 15, 16, 17, and 18 in Braun, Coleman, and Drew [5], as well as in Haberman [7]. Chemical kinetic equations are worked out in the first chapter of Murray [12].

Chapter Five

There is little of the general theory of partial differential equations that is needed in this chapter. A concise introduction, slanted towards applications in the biological sciences, is to be found in diverse chapters of Jones and Sleeman [8].

The algae bloom model in Section 5.2 is due to Kierstead and Slobodkin in 1953. Pollutant flow and oxygen uptake, modeled in Section 5.3, was first formulated by Streeter and Phelps in 1925 as part of a study on the self-purification of the Ohio River, and it is an early classic of environmental engineering.

An elementary discussion of traffic flow is given in Haberman [7]. A more thorough, but also more advanced, treatment that includes a discussion of shock waves is in Whitham: *Linear and Nonlinear Waves*, Wiley, 1974. See also the survey article by Lax, "Formation and Decay of Shock Waves", *Am. Math. Mo.* **79**, 1972, 227–241. Whitham's book contains additional information on Burger's equation as well.

Traveling wave solutions to Fisher's equation (Section 5.5) have an extensive literature dating back to the work of Kolmogorov, Petrovski, and Piskunov in 1937. See the books of Murray [12] and Jones and Sleeman [8] as well as the useful survey article of Hadeler, "Non-linear Diffusion Equations in Biology", in *Ordinary and Partial Differential Equations*, Springer Lecture Notes **564**, 1976.

Another area in which traveling waves occur is neurophysiology. A set of equations were derived by Hodgkin and Huxley in 1952. Simplified versions of their work were then developed by FitzHugh in 1961 and others and models the excitation of pulses in a neuron by means of an oscillator of the Van der Pol type. See the survey by Hastings, "Some Mathematical Problems From Neurophysiology", *Am. Math. Mo.* **82**, 1975, 881–895. A similar model for the excitation of heart muscles to produce contractions is discussed in detail in Chapter 7.

The growth of spatially inhomogeneous patterns is a topic whose mathematical treatment dates back to the work of A. Turing in 1952. Our version follows, in part, "Population Models in Heterogeneous Environments" by Levin which appeared in *MAA Studies in Math.* **16**, 1978, 439–476. Figure 5.16 is redrawn from Smith: *Mathematical Ideas in Biology*, Cambridge, 1968, which is also recommended for a fairly non-technical overview of other topics of interest.

The material in Section 5.6 on flows in channels is covered in an article by A. Ippen, "Tidal Dynamics in Estuaries", which appears in *Estuary and Coastline Hydrodynamics*, McGraw–Hill, 1966 (ed. by Ippen).

Chapter Six

The mass–spring system with a moving belt is treated in Stoker [13] and it dates back to the earlier work of Lord Rayleigh in his book on the theory of sound in 1877.

The Poincaré–Bendixson Theorem is proven in Hirsch and Smale [1], and the Hopf Bifurcation Theorem in $\mathbf{R}^2$ is proven in an appendix to Murray [12]. Our intuitive "proof" is borrowed from Borrelli and Coleman [6]. A direct computational proof is given by Loud in "Some Examples of Bifurcation," *Bifurcation Theory and Applications*, Springer Lecture Notes **1057**, 1984. An introduction to the theorem of Hopf in general is in the first chapter of Marsden and McCracken: *The Hopf*

Bifurcation and Its Applications, Springer–Verlag, 1976. The Hopf theorem in $\mathbf{R}^3$ depends on a reduction to the planar case using the center manifold theorem. A nice introduction to this is in the first chapter of the book *Applications of Center Manifold Theory* by Carr, Springer–Verlag, 1980.

The Van der Pol equation is, together with the Duffing equation $\ddot{z} + r\dot{z} - \beta z + z^3 = 0$ (Duffing, 1918), the prototypical nonlinear equation, much studied since the pioneering work of Van der Pol in 1927 (the Duffing paper is reprinted in the collection of articles edited by Bellman and Kalaba mentioned above under Chapter 3). Interesting modifications include the FitzHugh neurological model referred to earlier (see also Exercise 6.5.12) and the heart beat model of Zeeman which is taken up in Chapter 7.

The non-quadratic predator–prey model of Example 6.6 is discussed by R. May in *Model Ecosystems*, Princeton, 1973. In connection with this, see also the article by Rosenzweig, "Paradox of Enrichment", *Science* **171**, 1971, 385.

We follow Clark [10] in the fish harvesting model of Example 6.7. This is based on earlier work of Gordon in 1954 and Schaeffer in 1957. The idea that unrestricted access to a common resource can lead to overexploitation is discussed in a celebrated article by Hardin, "Tragedy of the Commons", *Science* **162**, 1968, 1243–1247.

The fascinating idea of species diversity, which was treated from the point of view of Turing's model in Chapter 5, is also suggested in another context by the work of Paine, "Food Web Complexity and Species Diversity", *Am. Nat.* **100**, 1966, 65–76. The notion is that diversity can come about from sufficient complexity in the food chain. This is explained in Exercise 6.5.11.

Our treatment of the flywheel governor follows Pontryagin: *Ordinary Differential Equations*, Addison–Wesley, 1962, as well as the book *Theory and Application of The Hopf Bifurcation*, Cambridge Press, 1981, by Hasard, Kazarinov, and Wan.

Chapter Seven

A discussion of gradient dynamics and the proof that the positive limiting sets of gradient systems are equilibria can be found in Hirsch and Smale [1].

Catastrophe Theory is identified with the very suggestive but difficult book of R. Thom, *Structural Stability and Morphogenesis*, Benjamin, 1975. A useful introduction to the subject is provided in the first two chapters of C. Zeeman, *Catastrophe Theory*, Addison–Wesley, 1977. The model of a pumping heart (first considered by Van der Pol in 1928) is also due to Zeeman and is reprinted as Chapter 3 of his book referenced above under the title "Differential Equations for the Heartbeat and Nerve Impulse".

The spruce budworm model follows Ludwig, Jones, and Holling, "Qualitative Analysis of Insect Outbreak Systems: The Spruce Budworm and Forest", *J. Animal Ecol.* **47**, 1978, 315–332. A wonderful rendering of their work, which is also a good undergraduate introduction to differential equation modeling, is by Tuchinsky, *Man in Competition with the Spruce Budworm*, Birkhäuser, 1981.

The catastrophe model of the earth's magnetic field is that of Chillingsworth and Furness: "Reversals of the Earth's Magnetic Field", *Dynamical Systems: Warwick 1974*, Springer Lecture Notes **468**, 1975. This follows earlier work of Bullard in 1955. In Chapter 8, an alternate model is proposed for the reversal phenomena.

The fishery collapse (Exercise 7.5.4), follows Clark [10] while algae blooms (Exercise 7.5.7) is suggested by work of Steele and Henderson, "A Simple Plankton Model", Am. Nat. **117**, 1981, 676–691.

Resilience in ecological models is an idea discussed by Holling, "Resilience and Stability of Ecological Systems", *Ann. Review of Ecol.* **4**, 1973, 1–23. The article by S. J. Gould, "Evolution: Explosion, Not Ascent", *N.Y. Times*, Jan. 22, 1978, suggests that punctuated evolution may be a catastrophe model.

Chapter Eight

The concepts of trapping region and volume contraction in $\mathbf{R}^3$ follow the treatment in the first chapter of Ruelle, *Statistical Mechanics And Dynamical Systems*, Duke Turbulence conference, Duke University, 1977.

The three species competition model is due to May and Leonard, "Nonlinear Aspects of Competition Between Three Species", *SIAM J.* **29**, 1975, 243–253. The dynamo model of the earth's magnet, due to Rikitake (1958), is based on Cook and Roberts, "The Rikitake Two-Disc Dynamo System", *Proc. Camb. Phil. Soc.* **68**, 1970, 547–569 (see also the survey article by Bullard: "The Disc Dynamo", *Am. Inst. of Physics Conf. Proc.* **46**, 1978, 373–389). E. Lorenz's model of convection flow is given a detailed introduction in the article by Ruelle cited above. A nice treatment of the Lorenz model is also in Borrelli and Coleman (6).

The cascade of bifurcation in the discrete logistic model is discussed by May, "Simple Mathematical Models with Very Complicated Dynamics", *Nature* **261**, 1976, 459–467. This topic, as well as symbolic dynamics, the technique used to study the random behavior of the discrete system when $a = 4$, is fully treated in the book by Devaney: *An Introduction to Chaotic Dynamical Systems*, Benjamin, 1986.

In Section 8.3, we use the fact that all numbers in the unit interval, except for a set of zero measure, have binary expansions in which zero and one appear with equal probability. This is discussed in Feller, *An Introduction to Probability Theory*, (Second Edition, Vol. 1, Ch. 8), Wiley, 1957.

Very readable general articles on chaotic phenomena are by Ruelle," "Strange Attractors", *Math. Intelligencer* **2**, 1980, 126–137, Hofstadter, Metamagical Themas column "Strange Attractors: Mathematical Patterns Delicately Poised Between Order And Chaos" in Scientific American 245, 1981 and, at a slightly more technical level, by J. Ford, "How Random is a Coin Toss?", *Physics Today*, April 1983, 1–8, and L. Kadanoff, "Roads to Chaos", *Physics Today*, Dec. 1983, 46–53. A provocative account suitable for general audiences is Prigogine and Stengers, *Order Out of Chaos*, Bantam Books, 1984.

There is a connection between chaos and the topic of fractals. See the interesting book of Mandlebrot, *The Fractal Geometry of Nature*, Freeman, 1983.

Chapter Nine

A number of classic papers in control theory are collected in Bellman and Kalaba's book referenced above in connection with Chapter 3.

The maximum principle is proven in the book by Pontryagin, *et al.*, *The Mathematical Theory of Optimal Processes*, Wiley, 1962, from which also comes Example 9.2. A simple proof in a special case is available in Luenberger [9]. Most of the examples given in the present chapter involve a simple restriction on the control u of the form $\alpha \leq u \leq \beta$, for given α and β. This often leads to bang–bang controls, a fact first noted by Bushaw in 1952 and elaborated on by LaSalle in 1959.

The optimization model for the inverted pendulum is a special case of the quadratic regulator problem and is based on the work of Kalman in 1960.

Optimal harvesting and the turnpike principles, as well as Exercise 9.4.5, are discussed in Clark [10] although similar ideas also occur in the economic literature. Early work along these lines for the exploitation of nonrenewable resources was by Hotelling in 1931.

The application to rocket flight follows that in Lietmann, *The Calculus of Variation and Optimal Control*, Plenum, 1981. Exercise 9.4.8 is taken from Goh, Vincent, Leitmann, "Optimal Control of a Predator–Prey System", *Math. Biosciences* **19**, 1974, 263–286.

Index